MAKING SPATIAL DECISIONS
USING GIS
A WORKBOOK

Kathryn Keranen ▪ Robert Kolvoord

Esri Press
REDLANDS | CALIFORNIA

Esri Press, 380 New York Street, Redlands, California 92373-8100

Copyright 2012 Esri

All rights reserved. Second edition 2012

16 15 14 13 12   1 2 3 4 5 6 7 8 9 10

Printed in the United States of America

Ask for Esri Press titles at your local bookstore or order by calling 800-447-9778, or shop online at www.esri.com/esripress. Outside the United States, contact your local Esri distributor or shop online at www.eurospanbookstore.com/Esri.

Esri Press titles are distributed to the trade by the following:

In North America:

Ingram Publisher Services
Toll-free telephone: 800-648-3104
Toll-free fax: 800-838-1149
E-mail: customerservice@ingrampublisherservices.com

In the United Kingdom, Europe, Middle East and Africa, Asia, and Australia:

Eurospan Group
3 Henrietta Street
London WC2E 8LU
United Kingdom

Telephone: 44(0) 1767 604972
Fax: 44(0) 1767 601640
E-mail: eurospan@turpin-distribution.com

To RGK for his driving ability, humor, and constant support.
—K.K.

To Phil and Louise Kolvoord for nurturing the interests of a boy
who fell in love with computers.
—R.K.

CONTENTS

PREFACE

We hope this book helps build a bridge between your classroom and the centers of government, law enforcement, research, and other real-world entities that rely on geographic information system (GIS) technology to make important decisions.

Making Spatial Decisions Using GIS: A Workbook, second edition, puts actual spatial data into your hands—or more precisely, your computer—to analyze, interpret, and apply to various scenarios, including disasters and emergencies. You will make the type of decisions that affect an agency, a community, or a nation. The projects in this book involve GIS map-making of course, but they also help you hone your critical thinking skills and pursue independent study. We have chosen scenarios we think are relevant, thought-provoking, and applicable to a broad range of studies, not just geography. We also think you will enjoy finding solutions to challenging problems.

Making Spatial Decisions Using GIS: A Workbook comes with a DVD that contains the GIS data, worksheets, and other documents you will need to complete the projects. The book also allows you to download ArcGIS Desktop 10.0 software. This trial software requires the Microsoft Windows Vista, Windows 7, Windows XP, Windows 2003 Server, or the Windows 2008 Server operating system. This book also provides information about online resources.

Making Spatial Decisions Using GIS: A Workbook is a college-level text that presumes you have some prior GIS knowledge. This second edition features an updated design, fewer step-by-step basic instructions, and more introductory information to add context to the exercises. The technical prerequisites are outlined in the introduction, which also explains how to use the book, organize your workflow, and evaluate your work product. The introduction includes a summary of all five modules.

We have provided worksheets for your convenience. They will help you follow the exercises by providing a place in which you can log answers to questions along the way and keep track of the work to be completed. The answers to the individual questions will help you prepare your analysis to answer the larger problems in each module. Don't lose sight of the forest (the overarching problem) by spending too much time looking at the trees (the individual questions). Your instructor has supplemental resources available to assist you with the projects in this book.

This book stems from our many decades of experience working with students and teachers at all levels. We hope it is an enjoyable learning experience for you!

Kathryn Keranen and Robert Kolvoord

ACKNOWLEDGMENTS

We would like to thank all those who helped make this book possible:

At Esri: Thanks to Peter Adams, Esri Press manager, and his staff, in particular Claudia Naber, acquisitions editor; Mike Kataoka, project editor; Riley Peake, cartographer, and the Esri Press production staff. Special thanks to Jack Dangermond, Esri founder and president.

We are especially grateful to Tom Casady, chief of the Lincoln, Nebraska, Police Department; Tom Conry, GIS manager for Fairfax County, Virginia; and the Houston, Texas, Police Department's public affairs office for providing data and other essential material.

We also want to acknowledge the wonderful students and teachers with whom we get to work and who contributed time and talent to this project.

INTRODUCTION

Geographic information system (GIS) technology is a powerful way to analyze spatial data and is used in many industries to support decision making. From reassigning police personnel on the beat to locating sites for new urban development, GIS is at the center of important decisions. In *Making Spatial Decisions Using GIS: A Workbook*, you will analyze GIS data and learn to make and support spatial decisions. This scenario-based book presents five modules, each with real-world situations designed to expand your GIS skills as you make decisions using actual GIS data. The projects involve the complicated, sometimes messy, spatially related issues professionals from many disciplines face each day. The scenarios are more than step-by-step recipes for you to learn new GIS skills; they are opportunities to develop your critical-thinking prowess. As is true in so many situations, there is more than one "right answer," and you will have to decide the best solution for the problem at hand.

How to use this book

The scenario-based problems in this book presume that you have prior experience using GIS and are able to perform basic tasks using ArcGIS software (we are more specific about our expectations in the "Prior GIS experience" section). Each of the five modules in this book follows the same format. Project 1 gives step-by-step instructions to explore a scenario. You answer questions and complete tables on a worksheet provided for you. You reach decisions to resolve the central problem and develop a presentation to share your solution. Project 2 provides a slightly different scenario and the requisite data without step-by-step directions. You apply what you learned in project 1. Finally, each module includes an "On your own" section that suggests different scenarios you can explore by locating and downloading data and reproducing the analysis from the guided projects.

The modules in this book were not designed to be done in any particular order. Each can stand alone. If the order is not assigned by your instructor, explore the scenarios in each module and choose the order in which to do them based on your needs and interests.

GIS workflow

One major difference between these modules and other GIS-based lessons you may have done is the focus on GIS workflow: documenting and being systematic about the problem-solving process. Following a consistent GIS workflow is an important part of becoming a GIS professional.

We recommend the following GIS workflow:

1. Define the problem or scenario.
2. Identify the deliverables (mostly maps) needed to support the decision.
3. Identify, collect, organize, and examine the data needed to address the problem.
4. Document your work:
   a. Create a process summary.
   b. Document your map.
   c. Set the environments.
5. Prepare your data.
6. Create a basemap or locational map.
7. Perform the geospatial analysis.
8. Produce the deliverables, draw conclusions, and present the results.

A more detailed explanation is in the "Workflow" section on page xviii.

Process summary

The process summary is particularly important because it serves as a record of the steps you took in the analysis. It also allows others to repeat the analysis and verify or validate your results. A process summary might look something like this for an analysis of agricultural land in Vermont. It will, of course, vary for each project:

Map document 1

1. Prepare a basemap of Vermont.
2. Symbolize counties in graduated color using the POP2000 field.
3. Label lakes and major rivers.
4. Prepare a presentation layout.
5. Save the map document.

Map document 2

1. Name the data frame percentage of agricultural land.
2. Add the land-cover raster.
3. Symbolize land-cover raster with unique values by land-cover type.
4. Measure/calculate the area of all land in Vermont.
5. Select agricultural land cover using the raster calculator.
6. Measure/calculate the area of agricultural land.
7. Calculate the percentage of agricultural land.
8. Prepare a presentation layout.
9. Save the map document.

The modules

Each module focuses on a different use of GIS for local-level decision making, from establishing evacuation routes during a hazardous material spill, to understanding the changing demographics of America's urban areas, to analyzing crime patterns in a city, to locating the best places for new development in an urban area.

Hazardous emergency decisions

Accidents, natural disasters, and terrorist acts all involve chaotic homeland security situations that require a coordinated response based on sound information. GIS, when applied to these emergencies, saves lives and property. This module puts you at the scene of two highway emergencies in which hazardous materials threaten a wide area. Your GIS analysis will aid first responders who must deal with evacuating and sheltering people, rerouting traffic, and providing for helicopter access. You will create buffers, analyze traffic patterns, and assess the suitability of school sites as emergency shelters. You will use ArcGIS Network Analyst to conduct point-to-point routing and calculate route directions.

Demographic decisions

Nearly half the world's population lives in cities; of the nineteen largest cities in 2000, only four are in industrialized nations. Thus, the study of urban demographics spans the globe. In fact, in the twentieth century, the number of city dwellers increased fourteen-fold worldwide. Demographic data allows you to study trends in population growth, aging, housing, income, education, and other factors that play a part in increased urbanization. This module focuses on Chicago, Illinois, and Washington, D.C., two metropolitan areas with complex demographic issues. You will analyze diversity indexes, examine 3D images, create histograms, and investigate housing values in this module. You will use the Spatial Statistics mean center and directional distribution to identify the geographic center and measure the trend for an area.

Law enforcement decisions

A geospatial approach to crime fighting helps decision makers deploy limited police resources—personnel, equipment, facilities—for maximum benefit. This module focuses on law enforcement in Houston, Texas, and Lincoln, Nebraska, two cities that have successfully incorporated GIS technology into their crime analysis and planning processes. You have the opportunity to use actual data to size up the crime situation in each city and recommend specific action plans based on your GIS analysis. You will work with buffer zones, geocoding, time analysis of crime, and mapping density. The maps you produce will be the type of effective visual representations that, in the real world, assist decision makers and inform citizens.

Hurricane damage decisions

In 2005, Hurricanes Katrina, Rita, and Wilma destroyed homes, businesses, infrastructure, and natural resources along the Gulf and Atlantic coasts. In the aftermath of the storms, federal, state, and local governments, service agencies, and the private sector responded by helping to rebuild the hurricane-ravaged areas and restore the local economies. GIS helped responders assess damage, monitor the weather, coordinate relief efforts, and track health hazards, among many other critical tasks, by providing relevant and readily available data, maps, and images. In this module, you will access some of the same data that guided critical decisions, such as funding and safety measures, in the wake of Hurricanes Katrina and Wilma. You will map elevations and bathymetry, analyze flooded areas and storm surges, and pinpoint vulnerable infrastructure. You will develop animations of the storm's path and parameters. This module features image classification to define clusters and output a classified image. You will also work with time-enabled feature classes and graphs to visualize multiple variables. In the real world, this process saves lives, time, and money.

Urban planning decisions

There is an abundance of spatial data to help pinpoint the *where*. Where should I buy a house? Where should I establish my business? GIS can help gather, analyze, and visualize data to create reliable location intelligence that results in sound decisions. In this module, you will apply similar GIS processes to help planners in large urban areas identify where to expand development. San Diego, California, and Seattle, Washington, are the laboratories for this module. You will identify decision criteria, develop distance layers, classify elevation, and use fuzzy logic and overlays to identify the ideal places for additional development. You will produce siting decision maps using the Fuzzy Membership and Fuzzy Overlay tools. These spatial skills can be applied to countless siting decisions, large and small.

Assessing your work

Your instructor will talk with you about assessment, but you can assess your own work before handing it in. The items below will help ensure your presentation maps are the best they can be. Think about each of these items as you finish your maps and write up your work.

Map composition

Do your maps have the following elements?
- title (addresses the major theme in your analysis)
- legend
- scale
- author (your name)

Classification

Did you make reasonable choices for the classifications of the different layers on your maps? Is the symbology appropriate for the various layers?

- For quantitative data, is there a logical progression from low to high values and are they clearly labeled?
- For qualitative data, did you make sure not to imply any ranking in your legend?

Scale and projection

- Is the map scale appropriate for your problem?
- Have you used an appropriate map projection?

Implied analysis

- Did you correctly interpret the color, pattern, and shape of your symbologies?
- Does any text you have written inform the reader of the map's intended use?

Design and aesthetics

- Are your maps visually balanced and attractive?
- Can you distinguish the various symbols for different layers in your maps?

Effectiveness of map

- How well do the map components communicate the story of your map?
- Do the map components take into account the interests and expertise of the intended audience?
- Are the map components of appropriate size?

By thinking about these items as you produce maps and do your analysis, you will make your maps the most effective they can be at solving the problems in each module.

Esri provides mapmaking novices with practical instruction on how to produce effective displays. Visit the Esri Mapping Center online at http://mappingcenter.esri.com/. This site is a central repository for "the use of ArcGIS in the graphic delivery of geographic information." It is a great resource for tips on cartography and using ArcGIS to make high quality maps.

Prior GIS experience

In these modules, we presume that you have used ArcGIS software before and that you can do the following tasks:

- Navigate and find data on local drives, on network drives, and on CDs and DVDs.
- Name files and save them in a known location.
- Use ArcMap to connect to a folder.

- Use ArcCatalog to preview a data layer and look at its metadata.
- Add data to ArcMap by dragging layers from ArcCatalog or using the Add Data button.
- Rearrange layers in the Table of Contents.
- Identify the Table of Contents and the Map windows in ArcMap and know the purpose of each.
- Use the following tools:
  - Identify
  - Zoom in
  - Zoom out
  - Full extent
  - Pan
  - Find
  - Measure
- Symbolize a layer by category or quantity.
- Open the attribute table for a data layer.
- Select features by attribute.
- Label features.
- Select features on a map and clear a selection.
- Work with tables.
- Make a basic layout with map elements.
- Use the drawing tools to place a graphic on the map.

If you need some review, there are many great resources available, including the GIS Tutorial series and *Getting to Know ArcGIS Desktop* from Esri Press, and ArcGIS Desktop Help online.

Setting up the software and data

The software and data to complete the exercises are provided with this book.

ArcGIS Desktop 10

A 180-day trial version of ArcGIS Desktop 10, ArcEditor license (single use) software can be downloaded at http://www.esri.com/180daytrial. Use the code printed on the inside back cover of this book to access the download site, and follow the on-screen instructions to download and register the software.

Please note that the 180-day trial of ArcGIS Desktop 10 is limited to one-time use only for each software workbook. In addition, the ArcGIS Desktop 10 software trial is applicable only to new, unused software workbooks. The software trial cannot be reused or reinstalled, nor can the time limit on the software trial be extended.

The ArcGIS Desktop 10 software installation includes three ArcGIS Desktop extension products used in this book: ArcGIS 3D Analyst, ArcGIS Network Analyst, and ArcGIS Spatial Analyst. ArcGIS 3D Analyst includes the ArcScene and ArcGlobe applications, which are used for three-dimensional visualization and exploration of geographic data. ArcGIS Network Analyst and ArcGIS Spatial Analyst provide tools for specialized analysis tasks.

All the help and resources you need to get up and running with your trial software are provided here through videos, instructions, and commonly asked questions: http://www.esri.com/evalhelp.

Using the data DVD

Refer to the installation guides at the back of the book for detailed system requirements and instructions on how to install the software data. The data license agreement is found at the back of the book and on the data DVD. If you do not feel comfortable installing programs on your computer, please be sure to ask your campus technology specialist for assistance.

The software and data need to be installed on the hard drive of all computers you will use to complete these modules. Installations on a computer network server may result in slow performance.

Metadata

Metadata (information about the data) is included for all of the GIS data provided on the Data and Resources DVD. The metadata includes a description of the data, where it came from, a definition for each attribute field, and other useful information. This metadata can be viewed in ArcCatalog.

In each project, one of the first things you will do is explore the metadata for the various layers.

Troubleshooting ArcGIS

Exercise instructions are written assuming the user interface and user preferences have the default settings. Unless you are working with a fresh installation of the software, however, chances are you will encounter some differences between the instructions and what you see on your screen. This is because ArcMap stores settings from a previous session. This could vary which toolbars are visible, where toolbars are located, the width of the table of contents, or whether or not the map scale changes when the window is resized.

WORKFLOW

Addressing and analyzing a problem using GIS requires a structured approach similar to the problem-solving techniques you have used in other disciplines with other tools. By using this approach, you will be certain to develop a solution that can be shared with and repeated by other GIS users, and you will be able to communicate your results successfully to a broad audience.

In this book, we use the following steps to define the GIS workflow or procedure. In each project, you will address each step, in order, as you explore the problems and develop a solution or arrive at a decision. This is not the only way to work through a geospatial problem, but it is widely used in practice. Here are the steps explored in detail in the following pages:

1. Define the problem or scenario
2. Identify the deliverables needed to support the decision
3. Identify, collect, organize, and examine the data needed to address the problem
4. Document your work
5. Prepare your data
6. Create a basemap or locational map
7. Perform the geospatial analysis
8. Produce the deliverables, draw conclusions, and present the results

1. Define the problem or scenario

This is perhaps the most difficult part of the entire process. You must define what issue you are trying to address from an often complicated sea of information and perhaps competing interests.

This book presents real-world scenarios to help you with this process. You should always try to focus on the core issue in any situation calling for geospatial analysis. What decision needs to be made? Who is going to make it? What do people need to know to make a rational decision?

One approach is to write down a short description of the problem, including the general scenario, the stakeholders, and the specific issues that need to be addressed. It also helps to think about what decisions will ultimately be made using the data. Of course, the scenarios in this book require geospatial analysis using ArcGIS software to solve problems. Other kinds of analyses may be included in these scenarios, but the focus will be on geospatial problem-solving.

2. Identify the deliverables needed to support the decision

After you have defined the problem, you need to think about what maps and other visualizations you will produce to help you analyze and solve the problem. These may include the following:

- maps
- charts
- tables of calculations
- written analysis

By envisioning these maps, you will be able to identify the data required for your analysis. You will also be able to determine if you have defined your problem in sufficient detail to develop a solution. Along with specifying your data requirements, your list of deliverables will also guide your analysis. These first steps of your workflow are very iterative. You may need to go through them a few times before you feel ready to begin your analysis. In many projects, they are also the most complicated steps.

3. Identify, collect, organize, and examine the data needed to address the problem

Once you have defined your problem and identified the deliverables, it is time to search for data. As a starting point, you should identify data for a basemap and data to solve the problem you have defined. In many instances, you will have data at hand that will allow you to pursue your analysis. These data may have been provided for you (as in the first two projects of each module) or it may be part of a collection of data where you are working. The data may come from the Esri Data & Maps collection of DVDs. However, in some instances you may need to get data from other sources, such as the Internet, or you may need to collect your own data. If you need to do some sort of field study to collect your data, you will want to be sure to develop a clear protocol for taking data and follow it consistently. You will also want to think carefully about the design of a database to hold your measurements.

In all cases, you should be careful to identify your data sources and make sure you have permission to use the data for your particular problem. Make sure all data layers have appropriate metadata that describes various aspects of the data, including the creator of the data, the map projection, and the attributes included. You will also want to know if you have vector or raster data layers and the accuracy or resolution of each layer.

As you collect the data for your analysis, we strongly recommend that you adopt a standard for how you organize and store these data. This organization will make your analysis much easier, and you will be able to quickly find different layers and share your work with others.

A standard directory structure looks like this:

Project folder that includes:

- Data folder containing all data layers
- Document folder containing all project documentation
- Project.mxd (this is the ArcGIS map document)

You must be able to both read and write to all folders within the project folder. You can place the project folder at a convenient location in the network structure in which you work. Each module shows you how to save your data with relative paths to make data sharing even easier.

When you obtain data from other sources, you will often want to take a quick peek at the data to make sure you understand what the data actually represent. Remember, these first steps of your workflow are very iterative. You may need to go through them a few times before you feel ready to begin your analysis. In some instances, you may need to adjust your problem definition to let you use the available data, or the deliverables may need to be modified to make the analysis possible.

4. Document your work

Documenting your project and creating a process summary is critical to keeping track of the various steps in your analysis. To document your project, go to the File menu and select Document Properties. In the dialog box that appears, you can enter some of the basic information about your project. A process summary is simply a text document that keeps track of the different steps you use in your analysis. Too often, documentation of GIS work is left to the end of the project or is not done at all. We encourage you to start your process summary early in each project and to keep up with it as you proceed.

5. Prepare your data

Accurate GIS analysis may call for changing the units in which measurements will be made. You will also want to know the units in which various quantities are measured; are the elevations in feet or meters above sea level? You may need to provide geographic references to certain quantities, such as properly locating addresses or adding GPS-based data to your map display. You will be guided through these steps in the different projects.

6. Create a basemap or locational map

Finally, you are ready to make maps and perform your geospatial analysis. Your first step in this process should always be to build a locational map or basemap that shows the area you are studying. A basemap will typically contain the major features of the area such as roads and streams, and it will help orient you geographically to the area and its features. This is a good practice when solving any geospatial problem as it will give you a sense of the scale of your study area and the different features that may dominate that area.

7. Perform the geospatial analysis

Now it is time to get down to the problem solving. In this step, you will apply the different geospatial tools to the data you have compiled. These tools include selecting by attribute or location, classifying, interpolating, map algebra, measuring area, or a wide variety of other techniques. The point of your analysis is to produce the deliverables that you specified above and allow you to develop a solution to the problem you defined. Often you will find that the first set of analysis tools may not provide results that help solve the problem and you will need to refine your analysis and try other tools or techniques. Even if you have been very careful and thorough in your planning to this point, geographic data never lose their ability to surprise you. As you work through the scenarios in this book, you will learn a variety of advanced analysis techniques.

8. Produce the deliverables, draw conclusions, and present the results

Finally, you are satisfied with your analysis and ready to complete your work. You will first need to complete your deliverables (as defined earlier in the process). This may mean making map layouts, graphs, charts, or tables. You will want to finish documenting your analysis process. Remember to keep the principles of good cartographic design in mind when you make your deliverables.

Esri provides practical instruction on how to produce effective displays. Visit the Esri Mapping Center online at http://mappingcenter.esri.com/. This site is a central repository for "the use of ArcGIS in the graphic delivery of geographic information." It is a great resource for tips on cartography and using ArcGIS to make high quality maps.

You will also need to write a report that states your conclusions and justifies them, using the deliverables you have produced. Always keep the audience in mind as you prepare to report your results. A technically savvy audience will have very different needs from a group of high-level decision makers. Remember to finish your process summary.

Congratulations! You have finished your project and used a GIS workflow that will lead to success.

MAKING SPATIAL DECISIONS USING GIS
A WORKBOOK

HAZARDOUS EMERGENCY DECISIONS

Introduction

Accidents, natural disasters, and terrorist acts all involve chaotic homeland security situations that require a coordinated response based on sound information. GIS, when applied to these emergencies, saves lives and property. The two scenarios in this module put you at the scene of highway emergencies in which hazardous materials threaten a wide area. Your GIS analysis will aid first responders who must deal with evacuating and sheltering people, rerouting traffic, and providing for helicopter access. You will create buffers, analyze traffic patterns, and assess the suitability of school sites as emergency shelters. You will use the ArcGIS Network Analyst extension in this module.

Scenarios in this module

- An explosive situation in Springfield, Virginia
- Skirting the spill in Mecklenburg County, North Carolina
- On your own

GIS software required

- ArcGIS Desktop 10 (ArcEditor)
- ArcGIS Network Analyst

Student worksheets

The student worksheet files can be found on the Data and Resources DVD.

Project 1: Springfield student sheet
- File name: Springfield_student_worksheet.doc
- Location: Project1_Springfield\Documents

Resource material: Material Safety Data Sheet for Black Powder
- File name: MSDS_BP.pdf
- Location: Project1_Springfield\Documents

Project 2: Mecklenburg student sheet
- File name: Mecklenburg_student_worksheet.doc
- Location: Project2_Mecklenburg\Documents

Resource material: Material Safety Data Sheet for Chlorine
- File name: MSDS_chlorine.pdf
- Location: Project2_Mecklenburg\Documents

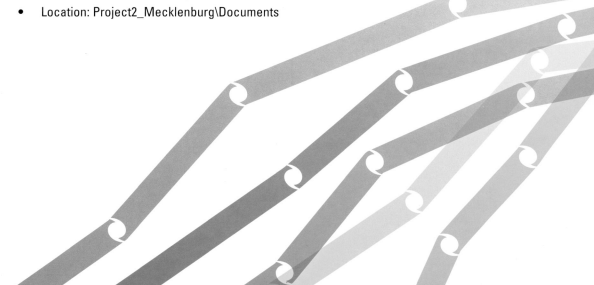

PROJECT 1

An explosive situation in Springfield, Virginia

Hazardous materials spills are a significant public health hazard and a major challenge for public safety officials who must respond to such life-threatening events. Hall and associates (1992) report on the consequences of hazardous materials spills across the United States. Good planning is critical when responding to such events. The *Hazardous Materials Spills Handbook* is the acknowledged reference for the details for almost every imaginable material (Bennett et al. 1982).

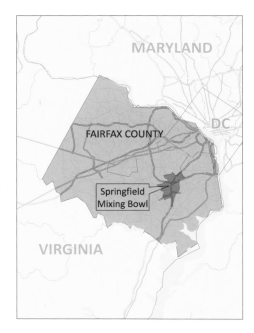

Scenario

Shortly before 4:00 AM on June 2, 1999, a tractor-trailer rig carrying 34,000 pounds of highly explosive black powder overturned at the "Mixing Bowl"—the heavily-traveled convergence of Interstates 95 and 495—in Springfield, Virginia. The flatbed tractor-trailer slid from the off-ramp from northbound I-95 onto westbound I-495. The Virginia State Police and the Fairfax County Police jointly provided personnel to coordinate the accident response.

4

Problem

Fairfax County police officers arrived first, immediately pinpointing the location with GPS receivers and identifying the hazardous substance. The officers accessed the Material Safety Data Sheet (MSDS) online for information about evacuation zones. An MSDS is required by the federal government and provides emergency personnel with proper procedures for handling or working with a particular hazardous substance. Officers needed maps showing the vulnerable area surrounding the accident, an estimate of the number of households to evacuate, suggestions for possible shelters for the evacuees, and a traffic analysis designating detours for vehicles. The analysis would include suggestions for a helicopter landing site both for medical evacuation and to transport personnel for logistical support. The helicopter landing site should be near the incident and evacuation zone.

MAKING
SPATIAL
DECISIONS
USING GIS

1

Hazardous
emergency
decisions

In applying GIS to a problem, you must have a very clear understanding of the situation. We find it helpful to answer these four questions that test your understanding and divide the problem into smaller problems that are easier to solve. Record your answers on the worksheet provided.

Q1 **What geographic area are you studying?**

Q2 **What decisions do you need to make?**

Q3 **What information would help you make the decisions?**

Q4 **Who are the key stakeholders for this issue?**
(This step is important. You need to know the audience for your analysis to help decide how to present your results.)

Deliverables

After identifying the problem, you need to envision the kinds of data displays (maps, graphs, and tables) that will provide the solution. We recommend the following deliverables for this exercise:

1. A map of Fairfax County showing roads and schools.
2. A map of buffered areas around the incident. The map should show the following:
   - Shelter locations
   - Residences to be evacuated
   - Helicopter landing site
3. Maps showing redirected traffic patterns both around the incident and within the buffer zone.

The questions in this project are both quantitative and qualitative. They identify key points that should be addressed in your analysis and presentation.

Tips and tools

Topical instructions are given in the following exercises. If more detailed instructions are needed, ArcGIS Desktop 10 provides these options:

MAKING

SPATIAL

DECISIONS

USING GIS

Hazardous

emergency

decisions

1. Use the Help file to ask a question or look up a keyword, such as a tab, menu option, or function. If you are online, it is better to use the Web-based help option by accessing the ArcGIS Resource Center for an up-to-date version of the help system included with the software.

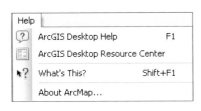

2. The tools can be accessed by using the traditional ArcToolbox or you can use the Search For Tools option found under the Geoprocessing menu.

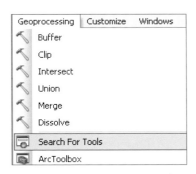

When you search for the tools, an explanation and a link to the tool appear.

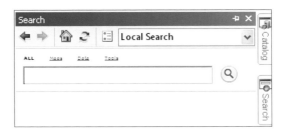

Examine the data

The next step in your workflow is to identify, collect, and examine the data for the Springfield accident analysis. Here, we have identified and collected the data layers you will need. Explore the data to better understand both the raster and vector feature classes in this exercise. In order to explore the data, you need to access the metadata associated with each feature class. Thoroughly investigate the data layers to understand how they will help you address the problem. The spatial coordinate system, the resolution of the data, and the attribute data are all important pieces of information about a feature class.

MAKING
SPATIAL
DECISIONS
USING GIS

1

*Hazardous
emergency
decisions*

1. Open ArcMap. (For these exercises, the Getting Started dialog box is not needed. Select the "Do not show this dialog in the future" option.)

2. Add Data by connecting to the folder \Project1_Springfield\Springfield_data.

   The data folder contains the Springfield geodatabase, which holds two feature datasets and an image. Add the image aerial and the features bldg, counties, and gschools from the feature dataset additional_layers.

There are different metadata styles that control how you view an item's description. The metadata data style "ISO 19139 Metadata Implementation Specification" supports formal metadata and allows the complete metadata to appear in the Item Description window. This setting must be made in the stand-alone ArcCatalog software.

3. Open ArcCatalog and under Customize choose ArcCatalog Options. Select the Metadata tab and then select "ISO 19139 Metadata Implementation Specification" from the Item Description menu.

4. Close ArcCatalog.

Q5 *View the item descriptions for these features and complete the following table on your worksheet.*

Layer	Publication Information: Who Created the Data?	Time Period Data Are Relevant	Spatial Horizontal Coordinate System	Data Type	Resolution for Rasters	Attribute Values
bldg	Fairfax County, VA	2007			N/A	Building attributes
aerial	U.S. Geological Survey	2002		Raster	0.3 meters	N/A

5. Close ArcMap.

7

Now that you have explored the available data, you are almost ready to begin your analysis. First you need to start a process summary, document your project, and set the project environments.

Organize and document your work

The following preliminary steps are essential to a successful GIS analysis.

MAKING
SPATIAL
DECISIONS
USING GIS

1

Hazardous
emergency
decisions

Examine the directory structure

The next phase in a GIS project is to carefully keep track of the data and your calculations. You will work with a number of different files and it is important to keep them organized so you can easily find them. The best way to do this is to have a folder for your project that contains a data folder. For this project, the folder named **\Project1_Springfield\Springfield_data** will be your project folder. Make sure that it is stored in a place where you have write access. You can store your data inside the results folder.

The results folder already contains an empty geodatabase named **\Springfield_results**. Save your map documents inside the **\Springfield_results** folder.

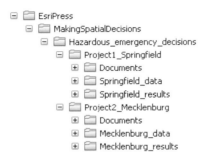

Create a process summary

The process summary is simply a list of the steps you used to do your analysis. We suggest using a simple text document for your process summary. Keep adding to it as you do your work to avoid forgetting any steps. The list below shows an example of the first few entries in a process summary:

1. Explore the data.
2. Produce a map of Fairfax County showing roads and schools.
3. Identify the incident.
4. Prepare a buffer zone around the incident.

Document the map

1. Open ArcMap and save the map document as incident1. **Save it in the** \Springfield_results **folder.**

You need to add descriptive properties to every map document you produce. Use the same descriptive properties for every map document in the module or individualize the documentation from map to map.

2. Access the Map Document Properties from the File menu. The Document Properties dialog box allows you to add a title, summary, description, author, credits, tags, and hyperlink base.

3. Select the Pathnames check box, which makes ArcMap store relative paths to all of your data sources. Storing relative paths allows ArcMap to automatically find all relevant data if you move your project folder to a new location or computer.

Set the environments

In GIS analysis, you will often get data from several sources and these data may be in different coordinate systems and/or map projections. When using GIS to perform area calculations, you would like your result to be in familiar units, such as miles or kilometers. Data in an unprojected geographic coordinate system have units of decimal degrees, which are difficult to interpret. Thus, your calculations will be more meaningful if all the feature classes involved are in the same map projection. Fortunately, ArcMap can do much of this work for you if you set certain environment variables and data frame properties. In this section, you will learn how to change these settings. To display your data correctly, you will need to set the coordinate system for the data frame. When you add data with a defined coordinate system, ArcMap will automatically set the data frame's projection to match the data. If you add subsequent layers that have coordinate systems different from the data frame, they are automatically projected on-the-fly to the data frame's coordinate system.

1. From the View menu, choose Data Frame Properties. Click the Coordinate System tab. Select Predefined, Projected Coordinate Systems, UTM, NAD 1983, Zone 18N, and click OK. This sets the projection for the active data frame.

2. From the Geoprocessing menu, choose Environment. Remember that the environment settings apply to all the functions within the project. The analysis environment includes the workspace where the results will be placed, and the extent, cell size, and coordinate system for the results.

3. Expand workspace. By default, inputs and outputs are placed in your current workspace, but you can redirect the output to another workspace, such as your results folder. **Set the Current Workspace as** \Project1_Springfield\Springfield_data\Springfield.gdb. **Set the Scratch Workspace as** \Project1_Springfield\Springfield_results\Springfield_results.gdb.

MAKING
SPATIAL
DECISIONS
USING GIS

1

Hazardous
emergency
decisions

4. For Output Coordinate System, select Same as Display.

5. Click OK and save the project again.

Analysis

Once you have examined the data, completed map documentation, and set the environments, you are ready to begin the analysis and to complete the displays you need to address the problem. A good place to start any GIS analysis is to produce a basemap to better understand the distribution of features in the geographic area you are studying. First, you will prepare a basemap of Springfield, Virginia, showing streets and schools.

STEP 1: Create a basemap of Fairfax County

1. Create a basemap of Fairfax County by adding counties and gschools from the Springfield geodatabase\Additional Layers.

2. From the Springfield geodatabase\Network Layers, add usastreets.

3. Symbolize the schools appropriately and import symbology from the "usa_street layer" file.

4. Locate the Capital Beltway (495), 95, and 395.

5. Use the label tool on the Draw toolbar to label Interstate 95 and Interstate 495. Interstate 495 is the Capital Beltway and it encircles Washington, D.C. Interstate 95 is the main north-to-south highway along the east coast. Both roads are heavily traveled.

6. Save the map document as incident1.

7. Save the map document again as incident2. (When you save the map document again as incident2 it correctly saves documentation, the data frame projection, and the environment settings. This saves you from redoing these variables for other deliverables.)

Deliverable 1: A map of Fairfax County showing roads and schools.

STEP 2: Identify the incident and isolate areas of concern

Upon arriving at the scene, police recorded the coordinates of the incident location as 77°10′33.031″W and 38°47′20.626″N. After questioning the driver of the tractor-trailer truck, officers learned that the black powder was not heavily encased. Knowing the exact nature of the material and how it was packed and stored allowed officers to use their MSDS database. Open the MSDS for black powder stored in the documents folder and determine the suggested extent of the evacuation area around the incident. The police designated an evacuation zone of 0.5 mile around the ramp where the incident occurred. The police team also designated shelters for evacuees within 0.5 mile of the evacuated area.

A. Create buffers around the incident

1. Open incident2 in ArcMap and be sure you are in Data View.

2. Select all elements from the Edit menu and delete the street labels.

3. Use the Go to the XY button on the Tools toolbar and locate the point where the incident occurred: 77°10′33.031″ W and 38°47′20.626″ N in degrees decimal minutes or −77.175842 and 38.789063 in decimal degrees. Click the Add Point button. Zoom to the point.

4. Use the Select Feature tool and select the ramp where the incident occurred. Right-click usastreets and select Data. Select Export Data and export the selection to \Springfield_results. gdb in the results folder. Name the file incident. (Note: The file must be saved as a File and Personal Geodatabase feature class.)

5. Clear the ramp selection and symbolize the incident layer so it can be easily seen.

6. Create buffers using the Multiple Ring Buffer tool:
   a. Set the Input feature as incident.
   b. Set the Output feature as buffer (and save it in the results folder).
   c. Distances are 0.5 and 1. (Type the number in the Distances box and then click the + button.)
   d. Make sure the Buffer Unit is Miles.

7. Zoom out so you can see the entire buffer.

8. Save the map document.

MAKING
SPATIAL
DECISIONS
USING GIS

1

*Hazardous
emergency
decisions*

B. Locate evacuation areas and shelters

You need to isolate schools that are within 0.5 to 1 mile from the accident scene. These schools are potential shelters for evacuees. Note that there are no schools within 0.5 mile of the incident, therefore no schools need to be evacuated.

1. Use the Clip tool to set the following parameters:
   a. The input feature is gschools.
   b. Clip Features should be set to buffer.
   c. Store the output feature in your results geodatabase and name it schools.
   d. Click Save and OK.

2. Remove gschools and symbolize the isolated schools. To decide how many people must be evacuated, you must isolate the residences within the 0.5-mile evacuation area.

3. Add bldg. Select the single family residential buildings from the bldg layer. (Hint: Query for Type = SFR. C stands for Commercial, I for Industrial, P for Public, and SFR stands for Single Family Residential.)

4. Select the 0.5-mile buffer area only.

5. Clip the bldg layer, name the file sfr, and store it in the results folder.

6. Remove bldg and clear the 0.5-mile buffer zone selection.

Q6   *How many single-family residences are in the danger area?*

Q7   *Where are the residential buildings located?*

Q8   *Which school would work best as a designated shelter for the people evacuated in the southeast quadrant?*

Q9   *Which school would work best as a designated shelter for the people evacuated in the southwest quadrant?*

7. Name the data frame buildings and shelters.

8. Save the map document.

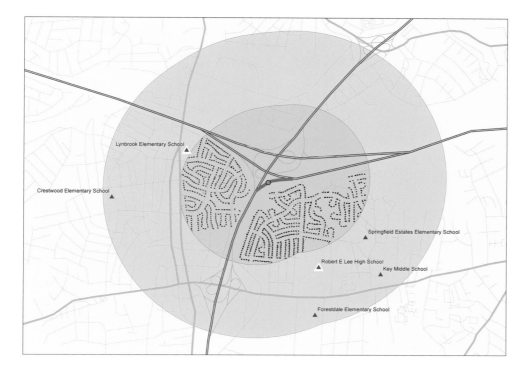

MAKING
SPATIAL
DECISIONS
USING GIS

1

Hazardous
emergency
decisions

C. Choose helicopter landing site

1. Right-click the buildings and shelters data frame and select Copy. Go to the Edit menu and select Paste.

2. Change the name of the pasted data frame to helicopter site.

3. Add the raster dataset aerial or, if you have access to the Internet, click Add Data from ArcGIS Online. Select World Imagery and high-resolution imagery will be added.

4. Make the 0.5-mile buffer hollow or turn on the Effects toolbar and swipe the buffer.

5. Examine the image closely by zooming in and out and pick a place that would be an appropriate helicopter landing site. It should be close to the incident for medical and other evacuations and for the logistical transport of emergency personnel.

6. Place a graphic where you think the landing site should be and symbolize it appropriately.

7. Save your map document as incident2.

8. Save the map document again as incident3.

Deliverable 2: A map of buffered areas around the incident. The map should show the following:
- Shelter locations
- Residences that need to be evacuated
- Helicopter landing site

MAKING
SPATIAL
DECISIONS
USING GIS

1

*Hazardous
emergency
decisions*

For the next part of the exercise you need to turn on the ArcGIS Network Analyst extension and the Network Analyst toolbar. You can do both of these tasks by using the Customize menu.

STEP 3: Map detours and best routes to shelters

The ArcGIS Network Analyst extension provides network-based spatial analysis including routing, travel directions, and address locations. The best source of street data that provides network analyst functionality is the Esri Data & Maps DVD. It includes StreetMap data with routing and address geocoding capabilities. These data can be used without any additional data preparation. To use these data, the user must clip out the study area and then build the network in ArcCatalog. For this exercise the network has been built for you.

After the immediate area is secure and the houses to be evacuated have been identified, the redirection of traffic is critical. In this part of the exercise you will determine alternative traffic routes and identify intersections requiring a police presence.

A. Map detours around the accident scene

1. Open incident3.

2. Remove the Helicopter Site data frame.

3. Add the usastreetsnet from the Network Layers feature class. When asked if you want to add all feature classes, click No. The Geographic Coordinate System Warning menu appears. Click Transformations, choose GCS_WGS_1984, click OK, and then click Close.

4. Zoom to the buffer layer.

5. On the Network Analyst toolbar, click the Show/Hide Network Analyst Window button. (Hint: It's the button to the right of Network Analyst.) The dockable Network Analyst window opens.

6. On the Network Analyst toolbar, click the Network Analyst menu and click New Route. The Network Analyst window now contains empty lists of Stops, Routes, Point Barriers, Line Barriers, and Polygon Barriers.

7. Right-click Stops (0) in the Network Analyst window and go to Load Locations. Load route1 from the Additional Layers feature class and click OK. When you load route1, stops 1 and 2 appear. These two stops are where road blocks will occur to block the highway. Using these stops, the software will calculate a path (route) around the closed road.

MAKING
SPATIAL
DECISIONS
USING GIS

1

Hazardous
emergency
decisions

Q10 **Route 1: Which road(s) is (are) being closed by these two stops?**

Every street that carries traffic within the 0.5-mile buffer zone needs to be closed. A barrier needs to be inserted at every street to keep traffic out.

8. Select the 0.5-mile buffer zone.

9. In the Network Analyst window, right-click Polygon Barriers and select Load Locations. Select the file buffer from the Springfield_results geodatabase. This places a barrier at all the roads leading into the 0.5-mile buffer zone. The 0.5-mile buffer must be selected.

10. Click the Solve button in the Network Analyst toolbar. Collapse the Route layer. A new alternative route is computed, avoiding the barriers. Directions are calculated for the alternative route and can be accessed by clicking the Directions window. Record the distance for each detour on the map.

You need to create two more alternative routes.

11. In the Network Analyst menu, click New Route.

12. Right-click Stops (0) in the Network Analyst window and go to Load Locations. Load route2 from the Additional Layers feature class.

Q11 **Route 2: Which road is being closed by these two stops?**

13. Select the 0.5-mile buffer zone. Right-click Polygon Barriers and select Load Locations. Select the file buffer. This places a barrier at all the roads leading into the 0.5-mile buffer zone. The 0.5-mile buffer must be selected.

14. Click the Solve button in the Network Analyst toolbar. Collapse Route 2 layer.

15

15. Repeat the above procedure for Route 3.

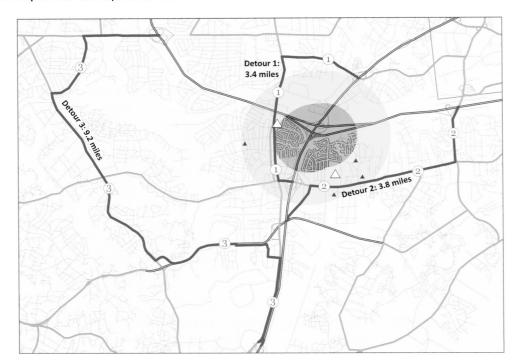

Q12   *Route 3: Which road is being closed by these two stops?*

16. Name the data frame detours.

17. Save the map document.

B. Determine best routes to shelters

1.   Copy the detours data frame.

2.   Paste it and rename the pasted data frame Best Routes.

3.   Collapse the detours data frame.

4.   Remove Route, Route2, and Route3. Clear the 0.5-mile buffer zone selection.

5.   Zoom to the southern quadrant of the buffer zone. This is the location of residential buildings that have to be evacuated. The southeast quadrant evacuation center is Robert E. Lee High School. The southwest quadrant evacuation center is Lynbrook Elementary School. The police want to establish the best route for evacuees so officers can direct traffic.

MAKING
SPATIAL
DECISIONS
USING GIS

1

*Hazardous
emergency
decisions*

6. In the southwest quadrant, the police are starting the evacuation route at 6780 Cabin John Rd., Springfield, VA, 22150, and ending at Lynbrook Elementary School. Prepare the best route so police can direct traffic.

7. In the Network Analyst toolbar, click the Network Analyst menu and click New Route.

8. Right-click Stops (0) and select Find Address.

9. Select Springfield from the data folder as the address locator.

10. Enter 6780 Cabin John Rd as the Street and click Find.

11. Right-click the address row to show the context menu and select Add Point. This puts a point at the 6780 Cabin John Rd. address.

12. Close the Find Menu.

13. In the Network Analyst toolbar, click the Create Network Location tool. Click the graphic point for the 6780 Cabin John Rd. address and then click Lynbrook Elementary School.

14. Click the Solve button in the Network Analyst toolbar to run the process and compute the best route.

There are two designated evacuation routes for the southeast evacuation area:

- The first route goes from 6015 Trailside Drive, Springfield, VA, 22150 to Robert E. Lee High School.
- The second route starts at 6756 Bison St., goes to 6700 Cimarron St., and ends at Robert E. Lee High School.

15. Repeat steps 11–14 for these two designated routes.

MAKING
SPATIAL
DECISIONS
USING GIS

1

Hazardous
emergency
decisions

16. Save the map document as incident3.

MAKING
SPATIAL
DECISIONS
USING GIS

1

*Hazardous
emergency
decisions*

Deliverable 3: Maps showing redirected traffic patterns both around the incident and within the buffer zone.

Once your analysis is complete, you still need to develop a solution to the original problem and present your results in a compelling way to the police in this particular situation. The presentation of your various data displays must explain what they show and how they help solve the problem.

Presentation

There are many ways, ranging from simple to advanced, that you can use to prepare a presentation. Whatever method of presentation you choose has to include a report documenting your analysis and addressing how you identified the evacuation shelters, road intersections to be closed, and the detour routes. You must explain the spatial patterns you see and describe the implications of your calculations and analysis for this problem. As you work through this book, try to use different types of presentation media. Pick the presentation medium that best fits your audience. Remember that your audience probably lacks your in-depth knowledge of GIS, so you will need to communicate your results in a way they will be able to understand and use.

Listed below are various presentation formats:

- Create a text document with inserted maps.
- Show your findings in a digitial slide presentation.
- Use ArcGIS Explorer Desktop, which is a free, downloadable GIS viewer that provides an easy way to explore, visualize, share, and present geographic information. This software can be downloaded at:

    http://www.esri.com/software/arcgis/explorer/download.html

MAKING
SPATIAL
DECISIONS
USING GIS

1

Hazardous
emergency
decisions

- Embed interactive maps in your text documents. These maps can also be shared with others. The links below provide information about this tool:

    http://www.esri.com/software/mapping_for_everyone/index.html
    http://help.arcgis.com/en/webapi/javascript/arcgis/index.html

- Create layer packages or map packages and share your maps with your classmates.
- Use ArcGIS Explorer Online to produce an interactive online geospatial presentation that can be shared: http://www.arcgis.com/home/

References

Bennet, G.G., F.S. Feates, and I. Wilder (1982). *Hazardous Materials Spill Handbook*. McGraw Hill, New York, NY.

Hall, H.I., G. S. Haugh, P. A. Price-Green, V. R. Dhara, and W. E. Kaye (1992). Risk factors for hazardous substance releases that result in injuries and evacuations: Data from nine states. *American Journal of Public Health*, 86(6), pp. 855–857.

PROJECT 2
Skirting the spill in Mecklenburg County, North Carolina

Scenario

Hazardous materials spills are a source of great concern for local and state law enforcement. In this hypothetical scenario, a tanker truck carrying chlorine gas was westbound on Interstate 85 north of Charlotte in Mecklenburg County, North Carolina. The driver lost control of his truck between the two segments of Interstate 77. The truck left the road and overturned. The impact caused several cracks in the tanker and gas began slowly leaking. The weather was cloudy with no wind.

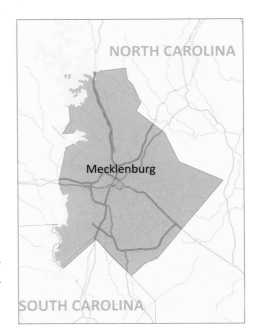

Problem

North Carolina state troopers arrived at the accident scene first, pinpointed the location with GPS receivers, and identified the leaking gas as chlorine. They then accessed the Material Safety Data Sheet (MSDS) for information about evacuation zones. (The MSDS for chlorine is provided in the documents folder.)

The state troopers immediately needed maps showing the required evacuation zone around the incident, a rough estimate of the population to evacuate, and possible shelter sites for evacuees. They also needed a traffic analysis to determine how best to reroute vehicles and where to pick up evacuees. The response team sought suggestions for a helicopter landing site, both for medical evacuation and to transport personnel for logistical support. The helicopter landing site needed to be close to the incident and the areas of evacuation.

Reminder: It helps to divide this large problem into a set of smaller tasks, such as the following:
1. Identify the geographic study area.
2. Determine the sequence of steps in your study.
3. Identify the decisions to be made.
4. Develop the information required to make decisions.
5. Identify stakeholders for this issue.

MAKING
SPATIAL
DECISIONS
USING GIS

2

*Hazardous
emergency
decisions*

The questions in this project are both quantitative and qualitative. They identify key points that should be addressed in your analysis and presentation.

Deliverables

We recommend the following deliverables for this exercise:
1. A map of Mecklenburg County showing roads and schools.
2. A map of buffered areas around the incident. The map should show the following:
   - Shelter locations
   - An approximate number of households to be evacuated
   - Helicopter landing site
3. Maps showing redirected traffic patterns both around the incident and from within the buffer zone.

Examine the data

The data for this project are stored in the **Project2_Mecklenburg\Mecklenburg_data** folder.

Reminder: View the item description to investigate the data. The following table can help you organize this information.

MAKING

SPATIAL

DECISIONS

USING GIS

2

Hazardous

emergency

decisions

Q1 *Investigate the metadata and complete the following table on your worksheet.*

Layer	Publication Information: Who Created the Data?	Time Period Data Are Relevant	Spatial Horizontal Coordinate System	Data Type	Resolution for Rasters	Attribute Values
blkgrp	Tele Atlas North America, Inc., Esri	2006			N/A	Demographic attributes
aerial	U.S. Geological Survey	2002		Raster	0.3 meters	N/A

Organize and document your work

Be sure to refer to the project 1, Springfield exercise and your process summary.

1. Set up the proper directory structure.
2. Create a process summary.
3. Document the map.
4. Set the environments:
   a. Set the Data Frame Coordinate System to be the same as the aerial raster dataset. You can do this by importing the coordinate system.
   b. Set the working directory.
   c. Set the scratch directory.
   d. Set the Output Coordinate System to Same as Display.

Analysis

An important first step in GIS analysis is to develop a basemap of your study area.

Complete deliverable 1 and answer the questions below to orient yourself to the study area.

Deliverable 1: A map of Mecklenburg County showing roads and schools.

Q2 *Which main road goes north to south?*

Q3 *Which main road goes west to east?*

To continue your analysis, you need to know the facts upon which the state troopers and first responders based their decision. The facts and timeline are as follows:

- The first North Carolina trooper on the scene reported the incident position as 80°50′38.76″W, 35°16′23.24″N (−80.8441, 35.273122).
- After talking to the trucker, troopers identified the leaking gas as chlorine.
- They contacted AccuWeather for a weather report. The weather for the rest of the day was cloudy with no wind.
- They carefully read the MSDS and decided to evacuate 2 miles around the incident and designate shelters that were within 0.5 mile of the outer perimeter of the evacuation zone.

Using all the information above, continue the analysis to prepare deliverable 2.

Reminder: When you create the multiple buffers, the distances should be 2 and 2.5 miles.

There is no building information with this exercise. You have to use the census block group data to estimate the number of households to evacuate within the 2-mile zone.

Clip the block group and then calculate statistics by using the households field to estimate the number of households to evacuate. It helps to show the block groups to be evacuated in graduated color by the number of households.

Q4 **Which quadrant has the most households to evacuate?**

Q5 **How many total households must be evacuated?**

Presentation

Keep in mind the interests and expertise of your audience as you prepare your presentation. Develop a solution to the original problem and present your results in a compelling way.

Refer to the list of presentation format options in project 1.

Extending the project

Your instructor may choose for you to complete this optional exercise.

Create reports on your layout with the names of the schools to be evacuated and the schools that can be used for shelters. Go to View/Reports and Create a Report.

MAKING
SPATIAL
DECISIONS
USING GIS

2

Hazardous
emergency
decisions

MAKING
SPATIAL
DECISIONS
USING GIS

2

Hazardous
emergency
decisions

Deliverable 2: A map of buffered areas around the incident. The map should show the following:

- Shelter locations
- An approximate number of households to be evacuated
- Helicopter landing site

The final deliverable consists of rerouting traffic both around the incident and from within the evacuation zone to shelters.

Deliverable 3: Maps showing redirected traffic patterns both around the incident and from within the buffer zone.

Refer to your process summary from project 1 of this module and review the procedure for creating barriers and saving them. Create detours 1–5 using the given detour stops layers (detour1stops, detour2stops, etc.).

Record the distance of each detour on the map.

Based on a data analysis, the following schools were designated as shelters. Several were not in the designated 0.5-mile area around the 2-mile buffer zone. There were designated pickup areas where buses were sent to evacuate the people with no other means of transportation. The pickup areas and associated shelter schools are listed below:

Hint: Some of the schools used for shelters are not in the 0.5-mile shelter zone. Add gschools from the Mecklenburg geodatabase to find the correct shelters.

Pickup area	Designated school shelter
3201 Graham St.	Highland Mill Elementary School
2201 Lasalle St.	Johnson C. Smith University
3601 Beatties Ford Rd.	Oakdale Elementary
4451 Statesville Rd.	Winding Springs Elementary

Create routes from the pickup areas to each of the shelters. (Hint: Use the usa_streets address locator in the data folder.)

Record the distance of each on the map.

MAKING

SPATIAL

DECISIONS

USING GIS

3

Hazardous

emergency

decisions

PROJECT 3

On your own

You have worked through a guided activity on the impact of a hazardous materials spill and repeated that analysis in another community. In this section, you will reinforce your skills by researching and analyzing a similar scenario entirely on your own.

First, you must identify your study area and acquire data for your analysis. There are many variations of this activity; consider choosing a scenario that has local impact. Think about chemicals that are transported through your community on railroads, pipelines, and highways. Identify locations such as nuclear power stations, oil refineries, fertilizer plants, gas stations, and dry cleaners that could be the source of hazardous spills. Don't forget nearby water resources. Rivers and lakes could be contaminated depending on where a spill occurs. There are many Web sites that contain extensive collections of Material Safety Data Sheets (MSDS). Make sure to access the appropriate MSDS for your hazardous material.

Refer to your process summary and the preceding module projects if you need help. Here are some basic steps to help you organize your work.

Research

Research the problem and answer the following questions:

1. What is the area of study?
2. What is the hazardous material and what are the MSDS constraints?

Obtain the data

Do you have access to baseline data? The Esri Data & Maps Media Kit provides many of the layers of data that are needed for project work. Be sure to pay particular attention to the source of data and get the latest version. Older versions of the Media Kit are very useful for temporal comparison, so be sure to check the date. For this exercise, it is imperative that you have access to Esri StreetMap data so you can use ArcGIS Network Analyst to calculate detours and evacuation routes. If you obtain data from your local GIS department, make sure to ask for a transportation network along with any other needed data. If you do not have access to the Esri Data & Maps Media Kit, you can obtain data from the following sources:

MAKING
SPATIAL
DECISIONS
USING GIS

3

*Hazardous
emergency
decisions*

- Census 2000 TIGER/Line Data http://www.esri.com/tiger
- Geospatial One Stop http://gos2.geodata.gov/wps/portal/gos
- The National Atlas http://www.nationalatlas.gov
- U.S. Geological Survey Seamless Data Warehouse http://seamless.usgs.gov (This site allows you to download a high-resolution image.)

Workflow

After researching the problem and obtaining the data, you should do the following:

1. Write a brief scenario.
2. State the problem.
3. Define the deliverables.
4. Examine the metadata or item description.
5. Set the directory structure, start your process summary, and document the map.
6. Decide what you need for the data frame coordinate system and the environments.
    a. What is the best projection for your work?
    b. Do you need to set a cell size or mask?
7. Start your analysis.
8. Prepare your presentation and deliverables.

Always remember to document your work in a process summary.

Presentation

Refer to the list of presentation format options in project 1.

DEMOGRAPHIC DECISIONS

Introduction

Nearly half the world's population currently lives in cities; of the nineteen largest cities in the year 2000, only four are in industrialized nations. Thus, the study of urban demographics now spans the globe. In fact, in the twentieth century, the number of city dwellers increased fourteen-fold worldwide.

Demographic data allows you to study population growth trends, aging, housing, income, education, and other factors that play a part in increased urbanization. This module focuses on Chicago and Washington, D.C., two metropolitan areas with complex demographic issues. You will analyze diversity indexes, examine 3D images, create histograms, calculate housing values, and identify mean centers and standard distances in this module.

Scenarios in this module

- For richer or poorer in Chicago
- Determining diversity in Washington, D.C.
- On your own

GIS software required

- ArcGIS Desktop 10 (ArcEditor)
- ArcGIS 3D Analyst

Other tools required

- Microsoft Excel
- Internet connection, preferably high-speed

Student worksheets

The student worksheet files can be found on the Data and Resources DVD.

Project 1: For richer or poorer in Chicago
- File name: Chicago_student_worksheet.doc
- Location: Project1_Chicago\Documents

Project 2: DC student sheet
- File name: DC_student_worksheet.doc
- Location: Project2_DC\Documents

PROJECT 1
For richer or poorer in Chicago

The history of relationships between different ethnic groups in the United States is fraught with tension and violence. The United States has long had a single dominant ethnic group, but this is changing and soon U.S. demographics will likely reflect no majority ethnic group. The history of the different ethnic groups informs their migration within the United States. For example, African-Americans live in significant numbers in the southeastern part of the country due to the initial settling of Africans in that region as slaves. After the Civil War, many chose to stay where they were, while a significant subset migrated to northern urban areas to take advantage of economic opportunities.

This migration leads to interesting questions:
- When ethnic minorities move to urban areas, do they remain in the same neighborhoods over time or do they spread uniformly or in some pattern throughout the metropolitan region?
- Once neighborhoods are segregated, do they remain so or does the segregation abate over time?
- What does this distribution say about the relationships and attitudes for relations between different ethnic groups?
- What influence do improving economic conditions have on this migration?

Sociologists have long studied these patterns using demographic data obtained from the U.S. Census to try to understand the change over time. In 2004, John R. Logan, Brian J. Stults, and Reynolds Farley published a study that explored the changes in segregation between African-Americans and Caucasians and found that such segregation had decreased slightly across U.S.

urban areas between the 1990 and 2000 census. They also found that segregation of Hispanics and Asians increased in that time period.

Q1   *Identify possible reasons for these changes in segregation.*

Scenario

In order to do their study, Logan, Stults, and Farley needed to investigate the census data and look for patterns in the 1990 and 2000 demographic data. In this activity, you will perform a similar investigation of Chicago to assess the distribution of different ethnic groups throughout that metropolitan area.

Problem

A local university's social science division needs maps and charts to explore the changing urban demographics of the Chicago metropolitan area and its relationship to Cook County over a ten-year period. A demographic research team is examining neighborhood integration and transition by measuring increasingly multiethnic and multiracial populations. The team would like to see maps of African-American, Caucasian, and Hispanic populations between 1990 and 2000. The team would also like to see diversity indexes as well as the mean centers and directional distribution of ethnic groups in Cook County calculated and displayed for both 1990 and 2000. African-Americans, Caucasians, and Hispanics represent the largest ethnic/racial groups in Chicago. Finally, the team wants to examine the impact of economics on diversity by studying the spatial distribution of median house value in 2000 and how that measure correlates to changing neighborhood settlement patterns. The researchers have requested that the data be displayed in 3D to engage their users. You must supply the researchers with a GIS-based spatial analysis using the data and tools introduced below.

In applying GIS to a problem, you must have a very clear understanding of the situation. We find it helpful to answer these four questions that test your understanding and divide the problem into smaller problems that are easier to solve. Record your answers on the worksheet provided.

Q2   *What geographic area are you studying?*

Q3   *What decisions do you need to make?*

MAKING
SPATIAL
DECISIONS
USING GIS

*Demographic
decisions*

(This step is important. You need to know the audience for your analysis to help decide how to present your results.)

MAKING
SPATIAL
DECISIONS
USING GIS

1

*Demographic
decisions*

Deliverables

After identifying the problem, you need to envision the kinds of data displays (maps, graphics, and tables) that will provide the solution. We recommend the following deliverables for this exercise:

1. A basemap showing Cook County and Chicago with census tracts from 2000. The map should show the population density classified in graduated color. Cook County and Chicago should be labeled. A written description of the spatial distribution of population in Cook County should be included with your map.

2. A series of maps for 1990 and 2000 with normalized population data for African-Americans, Hispanics, and Caucasians. A short written analysis of each ethnic group's spatial distribution should be included on the map layout.

3. Maps showing diversity indexes in 1990 and 2000. The percentage of African-American, Hispanic, Caucasian, and Asian populations should be shown with a bar graph.

4. A map showing the mean center and directional distribution of African-American and Hispanic population in 1990 and 2000. Add a short analysis of the changes in mean center and directional distribution between 1990 and 2000 on your map layout.

5. A 3D representation of the 2000 Diversity Index.

6. A distribution analysis of median house values in 2000 for Chicago and Cook County.

7. A double variable map of diversity index in relation to median house value with a written explanation.

8. An analysis of double variable maps of normalized African-American, Hispanic, or Caucasian data for 2000 shown in 3D in relation to median house value.

Tips and tools

Topical instructions are given in the following exercises. If more detailed instructions are needed, ArcGIS Desktop 10 provides these options:

MAKING

SPATIAL

DECISIONS

USING GIS

1

Demographic

decisions

1. Use the Help file to ask a question or look up a keyword, such as a tab, menu option, or function. If you are online, it is better to use the Web-based help option by accessing the ArcGIS Resource Center for an up-to-date version of the help system included with the software.

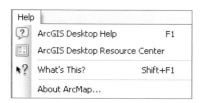

2. The tools can be accessed by using the traditional ArcToolbox or you can use the Search For Tools option found under the Geoprocessing menu.

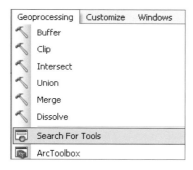

When you search for the tools, an explanation and a link to the tool appear.

The questions in this project are both quantitative and qualitative. They identify key points that should be addressed in your analysis and presentation.

Examine the data

The next step in your workflow is to identify, collect, and examine the data for urban analysis. Here, we have identified and collected the data layers you will need. Explore the data to better understand the vector feature classes in this exercise. In order to explore the data, you need to access the metadata associated with each feature class. Thoroughly investigate the data layers to understand how they will help you address the problem. The spatial coordinate system, the resolution of the data, and the attribute data are all important pieces of information about a feature class.

MAKING
SPATIAL
DECISIONS
USING GIS

1

*Demographic
decisions*

1. Open ArcMap. (For these exercises, the Getting Started dialog box is not needed. Select the "Do not show this dialog in the future" option.)

2. Add Data by connecting to the folder \Project1_Chicago\Chicago_data. **The data folder contains the Chicago.geodatabase, which holds four feature classes. Add the features Chicago, county, tracts_00, and tracts_90.**

There are different metadata styles that control how you view an item's description. The metadata data style "ISO 19139 Metadata Implementation Specification" supports formal metadata and allows the complete metadata to appear in the Item Description window. This has to be set in the stand-alone ArcCatalog software.

3. Open ArcCatalog and in the Customize menu choose ArcCatalog Options. Select the Metadata tab and select "ISO 19139 Metadata Implementation Specification" from the menu.

4. Close ArcCatalog.

The Microsoft Excel database has been downloaded from http://factfinder.census.gov and technical documentation can be found at that link. The database can be opened in Microsoft Excel to preview the available attributes.

Q6   ***View the item descriptions for these features and complete the following table on your worksheet.***

Layer	Publication Information: Who Created the Data?	Time Period Data Are Relevant	Spatial Horizontal Coordinate System	Data Type	Resolution for Rasters	Attribute Values
tracts_00		2006			N/A	
tracts_90	Esri Data & Maps 2000		Geographic		N/A	N/A

Q7 ***What information in tracts_00 and tracts_90 can be used to show possible residential patterns of African-Americans, Hispanics, and Caucasians?***

When you examine the data carefully, you will see that there are two populations given in tracts_00 and tracts_90. In tracts_00 the total population is given for 2000 and 2005. In tracts_90 the total population is given for 1990 and 1999. The 1999 and 2005 population values are only estimates of population and should not be used in calculations with the other attributes.

Q8 ***Is there information in tracts_00 that can be used to display median house value?***

5. Close ArcMap.

Now that you have explored the available data, you are almost ready to begin your analysis. First start a process summary, document your project, and set the project environments.

Organize and document your work

The following preliminary steps are essential to a successful GIS analysis.

Examine the directory structure

The next phase in a GIS project is to carefully keep track of the data and your calculations. You will work with a number of different files and it is important to keep them organized so you can easily find them. The best way to do this is to have a folder for your project that contains a data folder. For this project, the folder named **\Project1_Chicago\Chicago_data** will be your project folder. Make sure that it is stored in a place where you have write access. You can store your data inside a results folder. The results folder already contains an empty geodatabase named **\Chicago_results**. Save your map documents inside the **\Chicago_results** folder.

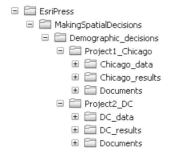

MAKING
SPATIAL
DECISIONS
USING GIS

1

Demographic

decisions

Create a process summary

The process summary is simply a list of the steps you used to do your analysis. We suggest using a simple text document for your process summary. Keep adding to it as you do your work to avoid forgetting any steps. The list below shows an example of the first few entries in a process summary:

1. Explore the data.
2. Produce a basemap showing Cook County and Chicago with census tracts from 2000.
3. Create maps for 1990 and 2000 with normalized population data for African-American, Hispanic, and Caucasian populations.

Document the map

1. **Open ArcMap and save the map document as** chicago1. **Save it in the** \Chicago_results **folder.**

You need to add descriptive properties to every map document you produce. Use the same descriptive properties for every map document in the module or individualize the documentation from map to map. You can access the Map Document Properties from the File menu. The Document Properties dialog box allows you to add a title, summary, description, author, credits, tags, and hyperlink base. After writing descriptive properties, be sure to select the Pathnames check box, which makes ArcMap store relative paths to all of your data sources. Storing relative paths allows ArcMap to automatically find all relevant data if you move your project folder to a new location or computer.

Set the environments

In GIS analysis, you will often get data from several sources and these data may be in different coordinate systems and/or map projections. When using GIS to perform area calculations, you would like your result to be in familiar units, such as miles or kilometers. Data in an unprojected geographic coordinate system have units of decimal degrees, which are difficult to interpret. Thus, your calculations will be more meaningful if all the feature classes involved are in the same map projection. Fortunately, ArcMap can do much of this work for you if you set certain environment variables and data frame properties. In this section, you will learn how to change these settings. To display your data correctly, you will need to set the coordinate system for the data frame. When you add data with a defined coordinate system, ArcMap will automatically set the data frame's projection to match the data. If you add subsequent layers that have a coordinate system different from the data frame, they are automatically projected on-the-fly to the data frame's coordinate system.

1. From the View menu, choose Data Frame Properties. Click the Coordinate System tab. Select Predefined, Projected Coordinate Systems, UTM, NAD 1983, Zone 16N, and click OK. This sets the projection for the active data frame.

2. From the Geoprocessing menu, choose Environment. Remember that the environment settings apply to all the functions within the project. The analysis environment includes the workspace where the results will be placed, and the extent, cell size, and coordinate system for the results.

3. Expand workspace. By default, inputs and outputs are placed in your current workspace, but you can redirect the output to another workspace such as your results folder. Set the Current Workspace as \Project1_Chicago\Chicago_data\Chicago.gdb. Set the Scratch Workspace as \Project1_Chicago\Chicago_results\Chicago_results.gdb.

4. For Output Coordinates select Same as Display.

5. Click OK and save the project again as chicago1.

MAKING
SPATIAL
DECISIONS
USING GIS

1

Demographic
decisions

Analysis

Once you have examined the data, completed map documentation, and set the environments, you are ready to begin the analysis and to complete the visualizations you need to address the problem. A good place to start any GIS analysis is to produce a basemap to better understand the distribution of features in the geographic area you're studying. First, you will prepare a basemap showing Chicago and Cook County. Include the 2000 census tracts displayed by population density on your basemap.

STEP 1: Create a basemap of Cook County

The basemap should show Cook County and Chicago with census tracts from 2000. The map should show the population density classified in graduated color. Cook County and Chicago should be labeled.

1. Add the tracts_00 and classify the data using graduated color. The POP00_SQMI field should be used as the Value field. Classify the data by manually altering the class breaks. Manual classification allows you to define your own classes. In this case you could use class breaks of 15,000; 30,000; 45,000; and 60,000. The number labels should be formatted to show 0 decimals.

2. Create a basemap showing Cook County population density and Chicago. Show the city limits of Chicago and label both Cook County and Chicago. Use proper cartographic principles.

37

3. Save the map document as chicago1.

4. Save the map document again as chicago2. (When you save the map document again as chicago2, it correctly saves documentation, the data frame projection, and the environment settings. This saves you from redoing these variables for other deliverables.)

Q9 *Describe the spatial distribution of population in Cook County.*

Deliverable 1: A basemap showing Cook County and Chicago with census tracts from 2000. The map should show the population density classified in graduated color. Cook County and Chicago should be labeled. A written description of the spatial distribution of population in Cook County should be included with your map.

STEP 2: **Compare race/ethnicity for 1990 and 2000**

To compare data, you need to create a standard classification that can be used for both datasets. This can be done by creating a layer definition and applying it to both the 1990 and 2000 data. To express values as a percentage you must normalize the data.

1. Open the map document chicago2 and remove tracts_00 and chicago.

2. Rename the data frame by clicking Layers and renaming the data frame Percentage African-American.

3. Add tracts_90 and classify the data using graduated color with Black (African-American) as the Value field and POP1990 as the normalization field. Classify the data by manually altering the class breaks. Manual classification allows you to define your own classes. In this case the normalized value cannot be greater than 1, which represents 100 percent. Suggested break values could be 0.2, 0.4, 0.6, 0.8, and 1.0. Format the labels as Percentage and save the layer file. It is important to note that the layer file that is to be used to symbolize all the data must include the maximum value possible in any of the layers. Save the layer file in the \chicago_results folder and give it an appropriate name, such as comparison.

4. Add tracts_00 and classify the data by using graduated color with Black (African-American) as the Value field and POP2000 as the normalization field. Import the symbology that you saved in step 3. Change the normalization field to POP2000.

5. Compare the data by using the Effects toolbar. Use Swipe or Flicker.

6. Copy the Percentage African-American data frame and paste. Rename the duplicate data frame Percentage Hispanics.

7. Make the Normalization field match either the 1990 or 2000 layer, respectively.

   Import the comparison layer file that you are using as the standard symbology and change the Value field from Black to Hispanic.

8. Repeat the process for Caucasians. (Hint: Use the data in the White field.)

9. Create a three-data-frame layout using correct cartography principles showing the percentage of African-Americans, Hispanics, and Caucasians in 2000. If you have access to the Internet, add data from ArcGIS Online. Select World Imagery and high-resolution imagery will be added. You can make the population layers partially transparent so you can see the underlying imagery by right-clicking a layer, selecting Properties, and in the Display tab, setting the Transparent value to 50%.

10. Save the map document as chicago2.

11. Save the map document again as chicago3.

MAKING
SPATIAL
DECISIONS
USING GIS

1

*Demographic
decisions*

Deliverable 2: A series of maps for 1990 and 2000 with normalized population data for African-Americans, Hispanics, and Caucasians. A short written analysis of spatial distribution for each ethnic group should be included on the map.

These questions may help focus your analysis:

Q10 **Why should you normalize the data?**

Q11 **How does the normalized data differ from the original data?**

Q12 **Describe the distribution of the African-American population and how it changed between 1990 and 2000.**

Q13 **Describe the distribution of the Hispanic population and how it changed between 1990 and 2000.**

Q14 **Describe the distribution of the Caucasian population and how it changed between 1990 and 2000.**

Q15 **Do you notice anything about the physical geography where there are large percentages of any particular ethnic group?**
   (Hint: Look for parks, bodies of water, open space, industrial areas, etc.)

MAKING
SPATIAL
DECISIONS
USING GIS

1

*Demographic
decisions*

STEP 3: Calculate and display the Cook County diversity index for 1999 and 2000

USA Today worked with college professors to develop a diversity index to represent racial and ethnic diversity with a single number. The first diversity index was created in 1991 and then updated for 2000. In this step, you will calculate the diversity index for 1990 and 2000, make observations about diversity in Cook County, and compare diversity in 1990 with that observed in 2000. Consult the following Web site for more background on the diversity index: http://www .unc.edu/~pmeyer/carstat/tools.html.

A. Calculate and study diversity index for 2000

1. Open the map document chicago3.

2. Remove two of the data frames. Remove all the files from the remaining data frame and rename it Diversity Index 2000.

3. Add tracts_00. Right-click tracts_00 and go to Data/Export data and export the data to chicago _results.gdb and name it tracts_00_DI. Remove tracts_00. When you export the file it is exported in the same projection as the display layer coordinate system (UTM, NAD 1983, Zone 16N).

4. Open the attribute table of tracts_00_DI. This attribute table is extremely long, and because you are dealing only with population, hiding the other fields is appropriate. You can hide a field by holding CTRL and double-clicking the field name. A hidden field will still be available in dialog boxes that list fields, but you can save space in the table window by hiding it. The following fields should remain visible: OBJECTID, Shape, FIPS, POP2000, WHITE, BLACK, AMERI_ES, ASIAN, HAWN_PI, OTHER, MULTI_RACE, and HISPANIC.

5. Save the map document.

6. In order to calculate the percentage of each attribute (WHITE, BLACK, AMERI_ES, ASIAN, HAWN_PI, OTHER, MULTI_RACE, and HISPANIC), you must add fields (Data Management) and make any necessary calculations. Percentage calculations will include decimals so we suggest that the field type be float. You will need to add the following fields for calculations:

    a. per_white   f. per_other
    b. per_black   g. per_hisp
    c. per_ameri_es   h. per_Nhisp
    d. per_asian   i. div_index
    e. per_hawnpi

7. Use the Field Calculator to calculate the percentage of each of the races and ethnic groups:
   a. Select by attributes WHITE>0.
   b. Using the Field Calculator, calculate the per_white field by using the formula WHITE/POP2000.
   c. If you examine the data, you can see that some of the field values are <NULL>. The <NULL> values need to be set to 0. You can select the <NULL> values by clicking the Switch Selection button and using the Field Calculator to enter 0 as the value.

8. Repeat the previous step for per_black, per_ameri_es, per_asian, per_hawnpi, per_other, and per_hisp, making sure to use the appropriate selection criteria.

   Step Summary:
   > Select by attribute WHITE>0
   > Use the Field Calculator: WHITE/POP2000
   > Switch Selection
   > Use the Field Calculator to enter 0 for the value <NULL>

MAKING
SPATIAL
DECISIONS
USING GIS

1

Demographic
decisions

9. For the field per_Nhisp, use the Field Calculator and enter 1 – [per_hisp]. Be sure to Clear Selected Features before performing the calculation.

10. To calculate the diversity index (div_index), use the Field Calculator and enter the following formula:
    1-([per_white] ^2 + [per_black] ^2 + [per_ameri_es] ^2+ [per_asian] ^2 + [per_hawnpi] ^2 + [per_other] ^2) * ([per_hisp] ^2 + [per_Nhisp] ^2) **(Note: MULTI_RACE is not included.)**

11. Display the div_index in graduated colors and exclude values of 1 or greater. Exclusion can be found under the Classify tab in the graduated colors menu or you can exclude values by using a **Definition Query** of div_index<1.

12. Create a layer file to use for comparing the diversity index in different years. The highest values are in the year 2000 so the values in this layer file would be the correct ones to use. Symbolize manually with four classes. Suggested values could be 0.2, 0.4, 0.6, and the highest value given. Format the labels. Higher values of the diversity index indicate greater ethnic diversity.

13. Save the layer file as div_index and save the map document (chicago3).

B. Calculate and study diversity index for 1990

The Census race/ethnic group definitions changed a bit between 1990 and 2000, so you will need to use a slightly different formula to calculate the 1990 diversity index. The rest of the process is the same.

1. Insert a new data frame and name it Diversity Index 1990. **Add tracts_90. Right-click tracts_90, go to Data/Export data, export the data to chicago_results.gdb, and name it** tracts_90_DI.

2. Add the following fields:
   a. per_white
   b. per_black
   c. per_ameri_es
   d. per_asian_pi
   e. per_other
   f. per_hisp
   g. per_Nhisp
   h. div_index

MAKING
SPATIAL
DECISIONS
USING GIS

1

*Demographic
decisions*

Due to the nature of the 1990 data—some of the census blocks have no reported population and the fact that you cannot divide by zero—you must slightly edit the data. For the 1990 data you must perform the following selections for the per_white field:

Step Summary:
Select by attribute POP1990>0 AND WHITE>0
Use the Field Calculator: WHITE/POP1990
Switch Selection
Use the Field Calculator to enter 0 for the value <NULL>

3. Repeat the previous step for per_black, per_ameri_es, per_asian, per_hawnpi, per_other, and per_hisp, with the appropriate selection.

4. For the field per_Nhisp, use the Field Calculator and enter 1 − [per_hisp]. Be sure to Clear Selected Features before performing the calculation.

5. To calculate the div_index, use the Field Calculator and enter the following formula:
   1-([per_white]^2+ [per_black]^2+ [per_ameri_es]^2+ [per_asian_pi]^2+ [per_other]^2) *
   ([per_hisp]^2+ [per_Nhisp]^2)

6. Symbolize div_index by importing the div_index layer file.

7. For comparison purposes, copy tracts_00_DI and paste in the Diversity Index 1990 Data Frame. (This will be removed later.)

8. Save the document (chicago3).

9. Compare the 2000 diversity index to the 1990 diversity index. You can do this by turning the layers on and off, or using the Effects toolbar and swiping or flickering the layers.

10. After making the comparison, remove tracts_00_DI from the Diversity Index 1990 Data Frame.

STEP 4: Examine diversity of individual census tracts

1. Insert a new data frame and name it Individual Census Tracts. **Copy the layer tracts_00_DI and paste it twice in the Individual Census Tracts data frame. Don't forget to set the coordinate system for the new data frame.**

2. **Display the top layer by Bar/Column with Whites, Blacks, Asians, and Hispanics normalized with POP2000. Make the background color hollow so you can see the 2000 classified diversity index layer as well.**

MAKING
SPATIAL
DECISIONS
USING GIS

1

*Demographic
decisions*

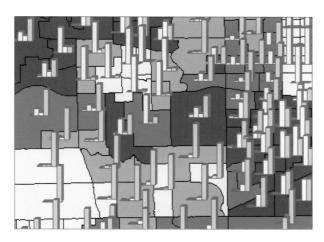

3. Save the map document as chicago3.

4. Save the map document again as chicago4.

Deliverable 3: Maps showing diversity indexes of 1990 and 2000. The percentage of African-American, Hispanic, Caucasian, and Asian population should be shown as a bar graph. (Hint: To make a layout showing multiple maps, you must insert a new data frame for each map you want to display.) In order to create deliverable 3, you must insert separate data frames for Diversity Index 1990, Diversity Index 2000, and Individual Census tracts. Include these maps in your written analysis.

These questions may help focus your analysis:

Q16   *Where are there clusters of high diversity?*

Q17   *What does low diversity mean?*

Q18   *Where are there clusters of low diversity?*

MAKING

SPATIAL

DECISIONS

USING GIS

1

Demographic

decisions

Q19 *In what ways did the diversity index change between 1990 and 2000?*

Q20 *Closely investigate Cook County by zooming in. What do you observe about the diversity of the census tracts?*

STEP 5: Calculate mean center and directional distribution

The mean center identifies the geographic center (or the center of concentration) for a set of features. It is useful for tracking changes in the distribution of ethnic groups over time.

Directional distribution measures the compactness of a distribution and shows the degree to which a distribution varies north to south (y-axis) and east to west (x-axis). In the following example, the size of the output ellipse encompasses 68 percent of the features (one standard deviation).

1. Open chicago4.

2. Remove the Individual Census Tracts data frame and the Diversity Index 2000 data frame.

3. Rename the **Diversity Index** Statistics 1990-2000.

4. Remove tracts_90 and add tracts_00_DI. You should now have tracts_90_DI and tracts_00_DI.

5. Calculate the mean center for the percentage of African-Americans for tracts_90_DI and tracts_00_DI. Give the features appropriate names such as mcb_90 and mcb_00 and store in Chicago_results.gdb. Use per_black as the Weight Field.

6. Calculate the directional distribution for the percentage of African-Americans (per_black) for tracts_90_DI and tracts_00_DI. Give the features appropriate names such as sdb_90 and sdb_00 and store in Chicago_results.gdb. Use 1 standard deviation for Ellipse Size and per_black as the Weight Field.

7. Repeat steps 5 and 6 using percentage Hispanic (per_hisp).

8. Label appropriately and show the mean center and directional distribution over time.

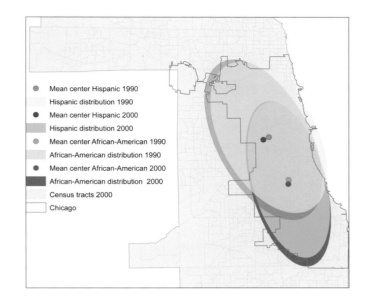

Legend:
- ● Mean center Hispanic 1990
- ☐ Hispanic distribution 1990
- ● Mean center Hispanic 2000
- ☐ Hispanic distribution 2000
- ● Mean center African-American 1990
- ☐ African-American distribution 1990
- ● Mean center African-American 2000
- ☐ African-American distribution 2000
- ☐ Census tracts 2000
- ☐ Chicago

MAKING
SPATIAL
DECISIONS
USING GIS

1

Demographic
decisions

9. Save the map document as chicago4.

10. Save the map document again as chicago5.

Deliverable 4: A map showing the mean center and standard distance of African-American and Hispanic population in 1990 and 2000. Add a short analysis of the changes in mean center and standard distance between 1990 and 2000 on your map layout.

These questions may help focus your analysis:

Q21 *Do the mean centers for African-American and Hispanic vary over time?*

Q22 *What would cause the mean centers to shift?*

Q23 *How did the mean centers shift between 1990 and 2000?*

Q24 *Do the directional distributions vary for each group between 1990 and 2000?*

Q25 *Why would directional distributions change?*

Q26 *What can you tell about changes in segregation from these maps?*

STEP 6: Display the data in 3D

ArcScene is a 3D visualization application that allows you to view your GIS data in three dimensions. The height of individual features is provided by either using base height (elevation) data or extruding features by particular attribute values.

You can control the vertical exaggeration, which allows you to exaggerate the difference in features to help you analyze variation in your data. While this technique is often very useful, you must not forget that ArcScene has exaggerated the view. In other words, it is possible to make Kansas look very mountainous with excessive vertical exaggeration, which might lead you to draw some very erroneous conclusions.

MAKING
SPATIAL
DECISIONS
USING GIS

1

Demographic
decisions

1. Open ArcScene, add tracts_00_DI, and import the div_index layer file from the results folder.

2. Refer back to "Organize and document your work" and do the following:
   a. Set the scene document properties. Don't forget to check "Store relative pathnames to data sources."
   b. Right-click Scene Layers and select Scene Properties. Click the Coordinate System tab and set the appropriate coordinate system.
   c. Set the environments.

3. Extrusion is the process of stretching a flat 2D shape vertically to create a 3D object. Use the ArcScene Extrusion tab and extrude tracts_00_DI by the attribute div_index. Each census tract gets a height proportional to its diversity index.

4. You can also control the vertical exaggeration to emphasize variations in height. Use the ArcScene Vertical Exaggeration tab (found under the Scene Properties General tab) and automatically calculate the vertical exaggeration based on the available data.

5. Save the map document as chicago_di.

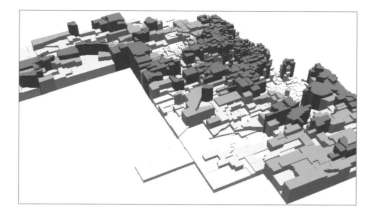

MAKING
SPATIAL
DECISIONS
USING GIS

1

Demographic
decisions

Deliverable 5: A 3D representation of the 2000 Diversity Index.

Q27 **What does looking at the data in 3D reveal?**

STEP 7: Analyze distribution and change in median house values

In this step, data preparation requires an Internet connection, an Internet browser, and Microsoft Excel.

Many factors can affect diversity in an urban setting, but economics is one of the most powerful forces. House values can serve as an economic indicator that can help you understand the changing patterns of diversity. Think about other indicators you might use to explore this situation.

You discovered earlier that tracts_00 did not have a median house value attribute. To compare median house value in 1990 and 2000, you must obtain these data from another source. You can download median house value data from http://factfinder.census.gov as a Microsoft Excel worksheet. The worksheet will have to be modified before it can be used in your geospatial analysis.

A. Download the data from U.S. Census Bureau

1. Using your Internet browser, go to http://factfinder.census.gov.

2. On the left side of the page, click DATA SETS.

3. Under the heading 2000, click the radio button beside Census 2000 Summary File 3 (SF 3) –Sample Data.

47

4. After checking the radio button, click Detailed Tables.

5. Under Select a geographic type, select State…County…Census Tract.

6. Under Select a state, choose Illinois.

7. Under Select a county, choose Cook County.

8. Under Select one or more geographic areas, choose All Census Tracts. Don't forget to click Add.

9. When all the census tracts populate the Current geography selections box, click Next.

10. The table that shows Median House value is H76, Median Value (Dollars) for Specified Owner-Occupied Housing Units. Choose that table and click Add.

11. Click Show Result.

12. Find the Print/Download tab at the top of the page. Click Download. Note: "Automatic prompting for file downloads" must be enabled on Internet Explorer to ensure a complete download of census tract data. For further assistance, click Download and click the link "Using FactFinder with Windows XP SP2."

13. In the window that appears, scroll down and click the radio button next to Microsoft Excel (.xls).

14. Click OK and save the downloaded file to the \Project1_Chicago\Chicago_results folder.

15. Unzip the data.

B. Clean up the data

1. Double-click dt_dec_2000_sf3_u_data1.xls and the spreadsheet will open in Excel. Verify that you have downloaded thousands of records, and not just a few.

2. To solve the missing median house value data in the tract_00 feature class, you require only two columns from the table. Delete all the other columns except the following:
   - 11- DIGIT Geography Identifier GEO_ID2, which is actually the FIPS number (Federal Information Processing Standard)
   - Specified owner_occupied housing units: Median value

3. Delete the first row.

4. Change the header of the Geography Identifier column to FIPS and the header of the Specified owner-occupied housing units: Median value to median_val. Change the Row Height of the first row to 15.

5. Save the spreadsheet to \Chicago_results as house_val. This file will be used for later analysis. In order to use this data, you need to join it to census tracts 2000.

C. Prepare the map

MAKING
SPATIAL
DECISIONS
USING GIS

1

Demographic
decisions

1. Open the map document chicago5.

2. Remove all layers and rename the data frame Median House Value.

3. Add tracts_00 and the spreadsheet file house_val.

4. Join the contents of the spreadsheet to the layer tracts_00. The common field between the spreadsheet and the layer is FIPS. Reminder: Join attributes from a table.

5. After the join you need to get rid of any of the tracts that did not join with the census data (NULL data) and you also need to get rid of any of the tracts that have 0 for a median house value. Open the joined attribute table and sort the median_val in ascending order. Select all the <NULL> and the 0 values. After you have selected these, switch the selection and export the file as tracts_00_2.

6. Remove house_val and tracts_00 from the table of contents.

7. Display tracts_00_2 by Graduated Color using the Value field median_val.

8. Reformat the legend and focus on the study area using proper cartography principles.

9. Save the map document as chicago6.

STEP 8: Create a histogram of median house values in 2000

1. Create a Histogram Graph using tracts_00_2 and the Graphs tool. Set the Value field to median_val.

2. Title and label the legends appropriately. Include the graph on the layout.

3. Save the map document as chicago5.

The graph is dynamically linked to the map. As you click a histogram bar, the census blocks with median house values in that range are highlighted on the map. Starting at the left (lowest house value), click each section of the histogram. Observe the spatial patterns on the map.

Deliverable 6: A distribution analysis of median house values in 2000 for Chicago and Cook County. Address the following questions in your map analysis:

Q28 What does the median value represent?

Q29 What is the lowest median_val?

Q30 What is the highest median_val?

Q31 Describe the spatial distribution of house values in Cook County.

Q32 What does the Count on the y_axis of the histogram represent?

Q33 What does the median_val on the x_axis represent?

Q34 Explain the patterns that you see in the graph.

STEP 9: Prepare a double variable map showing the relation between diversity index and median house value

1. Open ArcScene.

2. Refer back to "Organize and document your work" and do the following:
   a. Set the scene document properties. Select "Store relative pathnames to data sources."
   b. Set the data frame properties coordinate system.
   c. Set the environments.

3. Add tracts_00_DI and the spreadsheet house_val.

4. Join tracts_00_DI and house_val by the attribute FIPS.

5. Export the data and save the file as tractsdi_val. Remove tracts_00_di and the house_val.xls.

You now have one feature class with the diversity index and the median house value included. You do not have to save the project.

6. Display tractsdi_val by graduated color using the median_val field and exclude 0.

7. Extrude the feature class by the diversity index and Calculate from Extent the Vertical Exaggeration.

8. Save the map document as chicago_di_hval.

Q35 *Explain the double variable map.*

Deliverable 7: A double variable map of diversity index in relation to median house value with a written explanation.

MAKING
SPATIAL
DECISIONS
USING GIS

1

Demographic
decisions

STEP 10: Prepare 3D race/ethnicity maps relating to 2000 house values

1. Open ArcScene.

2. Add tracts_00_2.

3. Symbolize tracts_00_2 by importing the comparison.lyr file from step 2. Select Black and change the Normalization Field to POP2000.

4. Extrude by the attribute MEDIAN_VAL. Click the Calculate From Extent button to set the vertical exaggeration.

5. Repeat for Caucasian and Hispanic population percentages and observe their variations with respect to the extruded median house value.

6. Record your observations as you examine each of the scenes to help you in your analysis.

Deliverable 8: An analysis of double variable maps of normalized African-American, Hispanic, or Caucasian data for 2000 shown in 3D in relation to median house value.

Address the following questions in analyzing your map:

Q36 *Describe the relationship between median house value and the concentrations of different ethnic/racial groups.*

Q37 *What other variables might you analyze to understand the changing patterns of diversity in Chicago?*

MAKING
SPATIAL
DECISIONS
USING GIS

1

Demographic
decisions

Prepare a three-to-five-page summary analysis of your results for this problem. How has the distribution of minorities in the Chicago metropolitan area changed between the 1990 and 2000 Censuses? Can you predict what changes might be seen in the 2010 Census?

Once your analysis is complete, you still need to develop a solution to the original problem and present your results in a compelling way to the local university in this particular situation. The presentation of your various data displays must explain what they show and how they contribute to the overall analysis.

Presentation

There are many ways, ranging from simple to advanced, that you can use to prepare a presentation. Whatever method of presentation you choose has to include a report documenting your analysis and addressing the spatial patterns you observed. As you work through this book, try to use different types of presentation media. Pick the presentation medium that best fits your audience. Remember that your audience probably lacks your in-depth knowledge of GIS, so you will need to communicate your results in a way they will be able to understand and use.

Listed below are various presentation formats:
- Create a text document with inserted maps.
- Show your findings in a digital slide presentation.
- Use ArcGIS Explorer Desktop, which is a free, downloadable GIS viewer that provides an easy way to explore, visualize, share, and present geographic information. This software can be downloaded at http://www.esri.com/software/arcgis/explorer/download.html
- Embed interactive maps in your text documents. These maps can also be shared with others. The links below provide information about this tool:
    http://www.esri.com/software/mapping_for_everyone/index.html
    http://help.arcgis.com/en/webapi/javascript/arcgis/index.html
- Create layer packages or map packages and share your maps with your classmates.
- Use ArcGIS Explorer Online to produce an interactive online geospatial presentation that can be shared.
    http://www.arcgis.com/home/

Extending the project

Your instructor may assign you to complete this optional exercise to explore other demographic attributes.

There are many other demographic avenues to explore. Conduct an analysis of age groups (Age_65_up and Age_under5). Examine the percentage of males and females. Analyze vacant housing. Calculate the diversity index by school zones or congressional districts.

References

Fischer, C.S., G. Stockmayer, J. Stiles, and M. Hout (2004). Distinguishing the Geographic Levels and Social Dimensions of U.S. Metropolitan Segregation, 1960–2000. *Demography,* 41:1, pp. 37–59.

Logan, J., B.J. Stults, and R. Farley (2004). Segregations of Minorities in the Metropolis: Two Decades of Change. *Demography,* 41:1, pp. 1–22.

MAKING
SPATIAL
DECISIONS
USING GIS

1

Demographic
decisions

PROJECT 2

Determining diversity in Washington, D.C.

Scenario

Washington, D.C., is an economically and ethnically diverse community with interesting patterns of settlement and change. A public policy think tank is studying the changes in the nation's capital during the 1990s. It is interested in the spatial distribution of the residences of African-Americans, Caucasians, and Hispanics and the spatial distribution of median house value in the year 2000.

Problem

The think tank has contracted with you to make maps, charts, and graphs of residential patterns and median house value to use in its research.

Reminder: It helps to divide this large problem into a set of smaller tasks, such as the following:

1. Identify the geographic study area.
2. Determine the sequence of steps in your study.
3. Identify the decisions to be made.
4. Develop the information required to make decisions.
5. Identify stakeholders for this issue.

The questions in this project are both quantitative and qualitative. They identify key points that should be addressed in your analysis and presentation.

MAKING
SPATIAL
DECISIONS
USING GIS

2

Demographic

decisions

Deliverables

We recommend the following deliverables for this exercise:

1. A basemap showing Washington, D.C., with census tracts from 2000. The map should show the population density classified in graduated color. You can add the layers mjr_hwys and dtl_water to give you a basis from which to write descriptive narratives of the population density's spatial patterns.
2. A series of maps for 1990 and 2000 with normalized population data for African-American, Hispanic, and Caucasian residents displayed for each census tract. A short written analysis of spatial distribution of each ethnic group should be included on the map layout.
3. Maps showing diversity indexes of 1990 and 2000. The percentage of Caucasian, African-American, Asian, and Hispanic population should be shown as a bar graph.
4. A map showing the mean center and directional distribution of African-American and Hispanic population in 1990 and 2000.
5. A 3D representation of the 2000 Diversity Index.
6. A distribution analysis of median house values in 2000 for Washington, D.C.
7. A double variable map of diversity index in relation to median house value.
8. An analysis of double variable maps of normalized African-American, Caucasian, or Hispanic data for 2000 shown in 3D in relation to median house value.

Examine the data

The data for this project are stored in the **Project2_DC\DC_data** folder.

Reminder: View the item description to investigate the data. The table below helps you organize this information.

MAKING

SPATIAL

DECISIONS

USING GIS

2

Demographic

decisions

Q1 *Investigate the metadata and complete the following table on your worksheet.*

Layer	Publication Information: Who Created the Data?	Time Period Data Are Relevant	Spatial Horizontal Coordinate System	Data Type	Resolution for Rasters	Attribute Values
dc		2006		Vector	N/A	
tracts_90	Esri Data & Maps 2000		Geographic		N/A	

Organize and document your work

Be sure to refer to the project 1, Chicago exercise and your process summary.

1. Set up the proper directory structure.
2. Create a process summary.
3. Document the map
4. Set the environments:
   a. Set the Data Frame Properties Coordinate System to UTM, NAD 1983, Zone 18N.
   b. Set the working directory.
   c. Set the scratch directory.
   d. Set the Output Coordinate System to Same as Display.
5. 2000 Diversity Index = 1 − ([per_white]^2 + [per_black]^2 + [per_ameri_es]^2 + [per_asian]^2 + [per_hawnpi]^2 + [per_other]^2) * ([per_hisp]^2 + [per_Nhisp]^2) (Note: MULTI_RACE is not included.)
6. 1990 Diversity Index = 1 − ([per_white]^2 + [per_black]^2 + [per_ameri_es]^2 + [per_asian_pi]^2 + [per_other]^2) * ([per_hisp]^2 + [per_Nhisp]^2])

Analysis

An important first step in GIS analysis is to develop a basemap of your study area. Complete deliverable 1 and answer the question below to orient yourself to the study area.

Deliverable 1: A basemap showing Washington, D.C., with census tracts from 2000. The map should show the population density classified in graduated color. You can add the layers mjr_hwys and dtl_water to give you a basis from which to write descriptive narratives of the population density's spatial patterns.

Q2 ***Write a paragraph describing the spatial distribution of population in Washington, D.C.***

(The river on the western side of Washington, D.C., is the Potomac River. Constitution and Independence Avenues are the major streets that go west to east and meet at the Capitol. Georgia Avenue runs north to south.)

To complete deliverable 2, you need to make an appropriate legend. Create a layer file by using tracts_90 displayed in graduated color with whites as the Value field and normalizing by POP1990. The manual breaks should be set to 0.2, 0.4, 0.6, 0.8, and 1.0. Be sure to format the labels correctly and then use the layer file as the standard display scheme.

MAKING
SPATIAL
DECISIONS
USING GIS

2

Demographic
decisions

Deliverable 2: A series of maps for 1990 and 2000 with normalized population data for African-American, Hispanic, and Caucasian residents displayed for each census tract. A short written analysis of spatial distribution for each ethnic group should be included on the map layout.

Q3 ***Describe the changes in population for each race/ethnic group between 1990 and 2000.***

Deliverable 3: Maps showing diversity indexes of 1990 and 2000. The percentage of African-American, Hispanic, Caucasian, and Asian populations should be shown as a bar graph.

Q4 ***Where are there clusters of high diversity?***

Q5 ***Describe how the diversity index has changed from 1990 to 2000.***

Deliverable 4: A map showing the mean center and standard distance of African-American and Hispanic population in 1990 and 2000.

Q6 ***Do the mean centers for African-American and Hispanic vary over time?***

Q7 ***What would cause the mean centers to shift?***

Q8 ***How did the mean centers shift between 1990 and 2000?***

Q9 ***Do the directional distributions vary for each group between 1990 and 2000?***

Q10 ***Why would directional distributions change?***

Q11 ***What can you tell about changes in segregation from these maps?***

MAKING
SPATIAL
DECISIONS
USING GIS

2

Demographic
decisions

Deliverable 5: A 3D representation of the 2000 Diversity Index.

Q12 *Where are there clusters of high diversity?*

Q13 *Describe how the diversity index has changed from 1990 to 2000.*

Reminder: This deliverable requires ArcScene, the 3D viewing application. You need to extrude by the diversity index and then let the computer calculate the correct vertical exaggeration.

Before completing deliverable 6, you must download and clean up the census data for median house value of 2000. Refer to your process summary if you forget any of the steps.

Deliverable 6: A distribution analysis of median house values in 2000 for Washington, D.C.

Reminder: Save the file as **tracts_00_2** after you have joined the median house value data.

Q14 *Describe the spatial distribution of median house values in 2000.*

Reminder: Deliverables 7 and 8 require that you use ArcScene in the ArcGIS 3D Analyst extension. Set the vertical exaggeration after you extrude.

Deliverable 7: A double variable map of diversity index in relation to median house value.

Deliverable 8: An analysis of double variable maps of normalized African-American, Caucasian, or Hispanic data for 2000 shown in 3D in relation to median house value.

Presentation

Keep in mind the interests and expertise of your audience as you prepare your presentation. Develop a solution to the original problem and present your results in a compelling way.

Refer to the list of presentation format options in project 1.

Extending the project

Your instructor may choose for you to answer the following questions.

Q15 *How do the distributions of African-Americans, Caucasians, Hispanics, and median house values compare to the Chicago area?*

Q16 *How does the diversity index compare to Chicago?*

MAKING
SPATIAL
DECISIONS
USING GIS

3

*Demographic
decisions*

PROJECT 3
On your own

You have worked through a guided activity examining variation in the diversity index in an urban area, downloaded additional data, and repeated that analysis. In this section you will reinforce your skills by researching and analyzing a similar scenario, but entirely on your own. One of the first challenges is that you must identify your study area and acquire the data you need for your analysis. We suggest several possible demographic scales below. However, if there is another scale that has significant interest to you, download and work with that data.

- Urban areas of your state
- City boundaries
- School districts
- Congressional districts
- Counties

Refer to your process summary and the preceding module projects if you need help. Here are some basic steps to help you organize your work.

Research

Research the problem and answer the following questions before you begin:

1. What is the area of study?
2. What is the interest in this area and who are the stakeholders?

Obtain the data

Do you have access to baseline data? The Esri Data & Maps Media Kit provides many of the layers of data that are needed for project work. Be sure to pay particular attention to the source of data and get the latest version. Older versions of the media kit are very useful for temporal comparison—be sure to check the date.

MAKING
SPATIAL
DECISIONS
USING GIS

3

*Demographic
decisions*

If you do not have access to Esri Data & Maps, data can be obtained from the following sources:
- Census 2000 TIGER/Line Data http://www.esri.com/tiger—*Baseline data from the 1990 and 2000 Census can be found here.*
- Geospatial One Stop http://gos2.geodata.gov/wps/portal/gos
- The National Atlas http://www.nationalatlas.gov

Workflow

After researching the problem and obtaining the data, you should do the following:
1. Write a brief scenario.
2. State the problem.
3. Define the deliverables.
4. Examine the metadata or item description.
5. Set the directory structure, start your process summary, and document the map.
6. Decide what you need for the data frame coordinate system and the environments.
    a. What is the best projection for your work?
    b. Do you need to set a cell size or mask?
7. Start your analysis.
8. Prepare your presentation and deliverables.

Always remember to document your work in a process summary.

Presentation

Refer to the list of presentation format options in project 1.

LAW ENFORCEMENT DECISIONS

Introduction

A geospatial approach to crime fighting helps decision makers deploy limited police resources—personnel, equipment, facilities—for maximum benefit. In this module, we focus on law enforcement in Houston, Texas, and Lincoln, Nebraska, two cities that have successfully incorporated GIS technology into their crime analysis and planning processes. You have the opportunity to use actual data to size up the crime situation in each city and recommend specific action plans based on your GIS analysis. This module uses buffer zones, geocoding, and density mapping. The maps you produce will be the type of effective visual representations that, in the real world, assist decision makers and inform citizens.

Scenarios in this module

- Taking a bite out of Houston's crime
- Logging Lincoln's police activity
- On your own

GIS software required

- ArcGIS Desktop 10 (ArcEditor)
- ArcGIS Spatial Analyst

Other tools required

- Microsoft Excel
- Internet connection, preferably high speed

Student worksheets

The student worksheet files can be found on the Data and Resources DVD.

Project 1: Houston student sheet
- File name: Houston_student_worksheet.doc
- Location: Project1_Houston\Documents

Project 2: Lincoln student sheet
- File name: Lincoln_student_worksheet.doc
- Location: Project2_Lincoln\Documents

Resource material: UCR codes
- File name: UCR_codes.txt
- Location: Project2_Lincoln\Documents

1

PROJECT 1
Taking a bite out of Houston's crime

GIS offers law enforcement agencies (LEAs) a powerful tool to understand how reported crimes vary in both space and time. LEAs are increasingly adopting GIS as a tool to improve public safety and to better allocate their finite resources to better combat crime. Spatial data also help police explore various factors contributing to crime. GIS supports operational policing, tactical crime mapping, detection, and wider-ranging strategic analyses (Chainey and Ratcliffe 2005).

It is important to recognize that crime data have limitations and some criminologists and sociologists are concerned about the interpretation of such data using GIS (Hirschfield and Bowers 2001).

Scenario

In this exercise you will use data from the City of Houston to perform a crime analysis. The data come from the Positive Interaction Program (PIP), commonly called the PIP Crime Bulletin. Note that the data in this project are only for reported crimes and that the reporting process may result in some actual crimes not being "reported."

Problem

The Houston Police Department has the technology available to analyze reported crimes in order to more efficiently allocate patrol resources. You are a police department GIS analyst who has been directed to investigate the following:

- Crime patterns in proximity to police stations to determine if current patrols are effective or adjustments are necessary, such as reorganizing beats and/or building new substations.
- Patterns of aggravated assault, burglary, and narcotics-related crimes.
- Patterns of when crimes are committed.

MAKING
SPATIAL
DECISIONS
USING GIS

Law
enforcement
decisions

In applying GIS to a problem, you must have a very clear understanding of the situation. We find it helpful to answer these four questions that test your understanding and divide the problem into smaller problems that are easier to solve. Record your answers on the worksheet provided.

Q1 *What geographic area are you studying?*

Q2 *What decisions do you need to make?*

Q3 *What information would help you make the decisions?*

Q4 *Who are the key stakeholders for this issue?*
(This step is important. You need to know the audience for your analysis to help decide how to present your results.)

Deliverables

After identifying the problem, you need to envision the kinds of data displays (maps, graphs, and tables) that will provide the solution. We recommend the following deliverables for this exercise:

1. A basemap of Houston showing police stations, crimes, and roads. Include a graph of total crimes.
2. A map of police stations and locations of crimes along with recommendations for new police substation sites.
3. Density maps of aggravated assault, burglary, and narcotics-related crimes. Maps should have contour lines surrounding dense areas of crime, along with aerial or satellite imagery.
4. A graph exploring the time of day crimes are committed. Daily crime should be viewed temporally.

Tips and tools

Topical instructions are given in the following exercises. If more detailed instructions are needed, ArcGIS Desktop 10 provides these options:

1. Use the Help file to ask a question or look up a keyword, such as a tab, menu option, or function. If you are online, it is better to use the Web-based help option by accessing the ArcGIS Resource Center for an up-to-date version of the help system included with the software.

2. The tools can be accessed by using the traditional ArcToolbox or you can use the Search For Tools option found under the Geoprocessing menu.

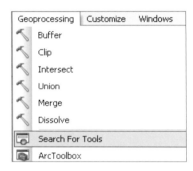

When you search for the tools, an explanation and a link to the tool appear.

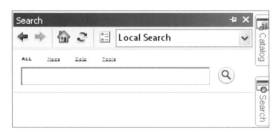

The questions in this project are both quantitative and qualitative. They identify key points that should be addressed in your analysis and presentation.

Examine the data

The next step in your workflow is to identify, collect, and examine the data for the Houston crime analysis. Here, we have identified and collected the data layers you will need. Explore the data to better understand both the raster and vector feature classes in this exercise. In order to explore the data, you need to access the metadata associated with each feature class. Thoroughly investigate the data layers to understand how they will help you address the problem. The spatial coordinate system, the resolution of the data, and the attribute data are all important pieces of information about a feature class.

1. Open ArcMap. (For these exercises, the Getting Started dialog box is not needed. Select the "Do not show this dialog in the future" option.)

2. Add Data by connecting to the folder \Project1_Houston\Houston_data. The data folder contains the Houston.geodatabase, which holds three feature classes and an Excel worksheet. Add the aug06.xls worksheet and the features blkgrp, Houston and usa_sts.

There are different metadata styles that control how you view an item's description. The metadata data style "ISO 19139 Metadata Implementation Specification" supports formal metadata and allows the complete metadata to appear in the Item Description window. This has to be set in the stand-alone ArcCatalog software.

3. Open ArcCatalog and in the Customize menu choose ArcCatalog Options. Select the Metadata tab and select "ISO 19139 Metadata Implementation Specification" from the Item Description menu.

4. Close ArcCatalog.

Q5 *View the item descriptions for these features and complete the following table on your worksheet.*

Layer	Publication Information: Who Created the Data?	Time Period Data Are Relevant	Spatial Horizontal Coordinate System	Data Type	Resolution for Rasters	Attribute Values
aug06.xls	http://www.houstontx.gov/police/stats.htm	2006	N/A	Excel worksheet	N/A	
blkgrp					N/A	N/A

5. Close ArcMap.

Now that you have explored the available data, you are almost ready to begin your analysis. First you need to start a process summary, document your project, and set the project environments.

Organize and document your work

The following preliminary steps are essential to a successful GIS analysis.

Examine the directory structure

MAKING
SPATIAL
DECISIONS
USING GIS

1

Law
enforcement
decisions

The next phase in a GIS project is to carefully keep track of the data and your calculations. You will work with a number of different files and it is important to keep them organized so you can easily find them. The best way to do this is to have a folder for your project that contains a data folder. For this project, the folder named **\Project1_Houston\Houston_data** will be your project folder. Make sure that it is stored in a place where you have write access. You can store your data inside the results folder. The results folder already contains an empty geodatabase named **\Houston_results**. Save your map documents in the **\Houston_results** folder.

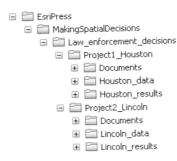

Create a process summary

The process summary is simply a list of the steps you used to do your analysis. We suggest a simple text document for your process summary. Keep adding to it as you do your work to avoid forgetting any steps. The list below shows an example of the first few entries in a process summary:

1. Explore the data.
2. Produce a basemap of the Houston area showing police stations, crimes, and roads. Include a graph of total crime.
3. Produce a map of police stations and locations of crimes.

Document the map

1. Open ArcMap and save the map document as crime1. Save it in the \Houston_results folder.

You need to add descriptive properties to every map document you produce. Use the same descriptive properties for every map document in the module or individualize the documentation from map to map. You can access the Map Document Properties from the File menu. The Document Properties dialog box allows you to add a title, summary, description, author, credits, tags, and hyperlink base. After writing descriptive properties, be sure to select the Pathnames check box, which makes ArcMap store relative paths to all of your data sources. Storing relative paths allows ArcMap to automatically find all relevant data if you move your project folder to a new location or computer.

Set the environments

In GIS analysis, you will often get data from several sources; these data may be in different coordinate systems and/or map projections. When using GIS to perform area calculations, you would like your result to be in familiar units, such as miles or kilometers. Data in an unprojected geographic coordinate system have units of decimal degrees, which are difficult to interpret. Thus, your calculations will be more meaningful if all the feature classes involved are in the same map projection. Fortunately, ArcMap can do much of this work for you if you set certain environment variables and data frame properties. In this section, you will learn how to change these settings. To display your data correctly, you will need to set the coordinate system for the data frame. When you add data with a defined coordinate system, ArcMap will automatically set the data frame's projection to match the data. If you add subsequent layers that have a coordinate system different from the data frame, they are automatically projected on-the-fly to the data frame's coordinate system.

1. From the View menu, choose Data Frame Properties. Click the Coordinate System tab. Select Predefined, Projected Coordinate Systems, UTM, NAD 1983, Zone 15N, and click OK. This sets the projection for the active data frame.

2. From the Geoprocessing menu, choose Environment. Remember that the environment settings apply to all the functions within the model. The analysis environment includes the workspace where the results will be placed, and the extent, cell size, and coordinate system for the results.

3. Expand workspace. By default, inputs and outputs are placed in your current workspace, but you can redirect the output to another workspace, such as your results folder. Set the Current Workspace as \Project1_Houston\Houston_data\Houston.gdb. Set the Scratch Workspace as \Project1_Houston\Houston_results\Houston_results.gdb.

4. For Output Coordinate System, select Same as Display.

5. For Extent, click the folder icon and select Houston from the Houston.gdb.

6. Expand the Raster Analysis Settings and set the Cell size to 200.

You also want to limit your analysis to the city of Houston. This is accomplished by using an analysis mask. The mask identifies those locations within the analysis extent that will be included when using a tool.

7. Set the Mask to Houston.

8. Click OK and save the project again as crime1.

MAKING
SPATIAL
DECISIONS
USING GIS

1

Law
enforcement
decisions

Analysis

Once you have examined the data, completed map documentation, and set the environments, you are ready to begin the analysis and to complete the displays you need to address the problem.

STEP 1: Geocode the data

Geocoding is the process of assigning a geographic location to an address. The accuracy of geocoding depends upon the accuracy of both the addresses you have and the reference street data used to find the geographic location.

To geocode data you need to build an address locator that pinpoints the location of the address on your maps. The address locator can be built by using:
- An online service (North America Geocode Service) provided by Esri
- An address locator style that supports the address attributes that are being used

A. Geocode police stations by using the online service (North America Geocode Service).

1. Add the following feature classes from \Project1_Houston\Houston_data:
   - Police_stations.txt
   - Houston
   - Usa_sts

2. Right-click police_stations.txt and select Geocode Addresses. Choose North America Geocode Service as the address locator.

3. Choose street as the Address Input Field.

4. Store the file as a feature class inside the \Houston_results.gdb.

5. Name the file police_stations.

6. Remove police_stations.txt.

Q6 **Were there any unmatched addresses?**

B. Geocode crime using the Dual Range Address Locator Style

1. Open aug06 in Microsoft Excel. There are several things that need to be fixed in order for this file to be geocoded:
- The block number and street name need to be combined into one field.
- The city and state names need to be added to the data.
- Many of the addresses are the same. This is because the data has been "sanitized" to protect the privacy of victims. The block number is given instead of a specific house number.

2. To combine the number and street name:
  a. Create a new field called Address.
  b. In the first cell under the Address cell, type the formula =CONCATENATE(F2, " ", G2).
  c. Highlight the cell.
  d. Drag the formula to the last record so that it calculates for all cells in the column.
  e. The block and street name should now be combined.

3. Add a field and name it city. **Copy and paste** Houston **in all cells.**

4. Add a field and name it state. **Copy and paste** TX **in all cells.**

5. Save this file as aug06_2. **(Save it in the** Houston_results **folder.)**

6. Create address locator:
  a. Use the US Address – Dual Ranges style.
  b. Reference data is usa_sts.
  c. Save the address locator as Houston in \Houston_results\Houston_results.gdb.

7. Add the file aug06_2. Double-click au6_2 and add the "click here for August 2006_Crim$" table.

8. Right-click August 2006_Crim$ and geocode addresses:
  a. Choose Houston as the Address Locator.
  b. Name the file raw_crime **and save it in** \Houston_results\Houston_results.gdb.

Q7 **What percentage of the addresses was unmatched?**

Q8 **How many total crimes were there?**

MAKING
SPATIAL
DECISIONS
USING GIS

*Law
enforcement
decisions*

MAKING
SPATIAL
DECISIONS
USING GIS

1

Law
enforcement
decisions

Q9   *What percentage of the crimes was matched? Tied?*

9. You need to keep only the crimes with addresses that were matched in your file. Go to Selection by Attribute and choose Score>0.

10. Export Geocoding Results and save it as a geodatabase feature class in the \Houston_results.gdb folder as Crime_match. Click the radio button for Use the same coordinate system as: the data frame.

11. Remove the Geocoding result: geocoded_crime.

12. Remove the click_here_for_August_2006_Crim$ table.

13. Save the map document as crime1.

STEP 2: Produce a basemap of Houston

1. Symbolize the police stations. (For this exercise you should use the More Symbols option in the Symbol Selector window and add the symbols for "Crime Analysis.")

2. Import the usa_sts.lyr layer. Accept CLASS_RTE as the Value field.

3. Check the crime_match data and make sure all the crimes are from August 2006. If you open the attribute table for crime_match and sort in ascending order by Offense_Date, you will see records that are not from the month of August 2006. Select all the files that are not from August 2006. With those records selected, select Switch Selection.

Q10   *How many records are not from August 2006?*

4. Export this selection and save the feature class as crime to eliminate the extraneous records. Remove crime.

5. Open the attribute table for crime and carefully study the data.

Q11   *How many crime records are listed?*

Q12   *Can multiple record (crime incidents) occur at the same location? Give an example.*

Q13 *Can you suggest a reason for this?*

Q14 *How is the Offense_Time represented?*

Q15 *What are the offenses recorded?*

MAKING
SPATIAL
DECISIONS
USING GIS

*Law
enforcement
decisions*

For additional information about Texas Criminal Codes, go to http://www.sadwilawyer.com/ texas%20law.htm.

6. Open the attribute table for crime, right-click Offense, and select Summarize to summarize the crimes.

7. Name the summary file total_crime and save it as a dBASE table in your results folder geodatabase. Add the file to the table of contents.

8. Open the total_crime attribute table. Turn on the Editor toolbar and start editing total_crime in the \Houston_results.gdb **folder. Change the offense labels as follows:**

Original Name	New Name
Aggravated Assault	Assault
Auto Theft	Auto Theft
Burglary	Burglary
Burglary of a Motor Vehicle	Burglary/Auto
Narcotic Drug Laws	Drugs
Driving While Intoxicated	DWI
Murder & Nonnegligent Manslaughter	Murder
Forcible Rape	Rape
Robbery	Robbery

9. Stop editing and save.

10. Create a graph from a table using the total_crime table. The graph should have the following properties:
- Vertical Bar Graph type with Cnt_Offense as the Value field.
- Select Cnt_Offense and Ascending for the x field.
- Label the x field with Offense.
- Enter an appropriate title, select a palette, and view in 3D.

11. Save the map document as crime1.

12. Save the map document again as crime2. (When you save the map document again as crime2, it correctly saves documentation, the data frame projection, and the environment settings. This saves you from redoing these variables for other deliverables.)

Deliverable 1: A basemap of Houston showing police stations, crimes, roads, and census blocks. Include a graph of total crimes.

MAKING
SPATIAL
DECISIONS
USING GIS

1

*Law
enforcement
decisions*

STEP 3: Produce a map of police stations with crime proximity

There are several ways to explore the spatial prevalence of crime in Houston. You will use two different methods to calculate the proximity of crime to the police stations. First, you will create buffers of fixed distance to see how many crimes occur within 2, 4, or 6 miles of the police stations. Then, you will use the technique of spatial join to connect each crime to the nearest police station.

1. Open crime2, delete the graph, and remove total_crime.

To allocate personnel and project where to build new substations, the police need to know where personnel resources are strained.

2. Create a multiple ring buffer with police stations as the input feature with buffer units set to miles and the distances as 2, 4, and 6. Name the file buffer and save in \Houston_results.gdb.

Now that there are buffer zones around the police stations, perform a spatial join with crime to connect each crime to the 2, 4, or 6 mile buffer. A spatial join appends one feature class's attribute table to another based on the relative locations of the features in the two layers. To do a spatial join, you need each crime to count as one event.

3. Add field (Data Management) to crime. Crime is the Input Table, the Field Name is Event, and the Field Type is SHORT.

4. Use the Field Calculator to perform the following operation: Event = 1.

5. Right-click buffer and select Joins and Relates. Select join to connect each crime to the appropriate buffer. From the menu, choose "Join data from another layer based on spatial location" and then choose the crime layer. Select Sum as the way you want the attributes to be summarized. This will spatially join the crimes to the appropriate buffer so that you will know how many crimes occur within the specified distance from each police station. Name the output feature buf_crime.

6. Symbolize buf_crime by unique value with Count as the Value field.

7. Remove buffer from the table of contents.

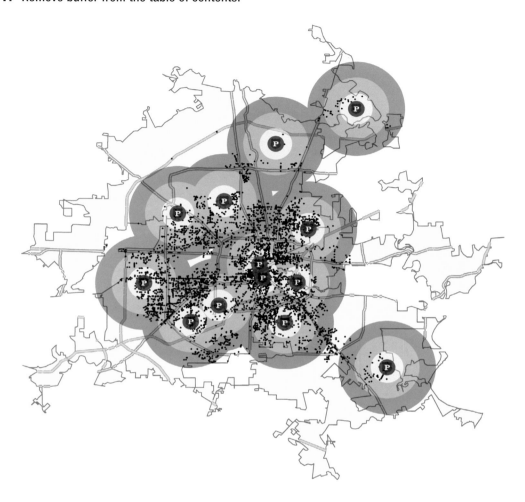

MAKING
SPATIAL
DECISIONS
USING GIS

1

*Law
enforcement
decisions*

Q16 *Complete this table on your worksheet.*

(Hint: Use the Sum_event field.) Remember, there are 7,439 crimes.

Distance in Miles	Crimes	Percent
2		
4		
6		
More than 6		

Q17 What do these results tell the police department about the distribution of crime?

The preceding table shows one method to categorize the crime by location using distance buffers.

Another way to characterize the crime distribution is to do a spatial join of all the crimes with the nearest police station. When you spatially join crimes to police stations, ArcGIS gives a summary of the numeric attributes of the features in the layer being joined (crimes) that are CLOSEST to each feature in the base layer (police_stations) and a count field showing how many crimes are closest to each station.

8. Repeat the process in step 5 for joining police stations to crime.

9. Name the output layer pol_sta_crimes.

Q18 Use the attribute table of pol_sta_crimes to complete this table on your worksheet.

Remember, there are 7,439 crimes.

Police Station	Number of Crimes	Percent
Central		
Clear Lake		
East (Magnolia)		
Fondren		
HPD Headquarters		
Intercontinental Airport		
Kingwood		
North		
Northeast		
Northwest		
South Central		
Southeast		
Southwest		
Westside		

10. Symbolize the number of crimes for each station as a bar chart by using sum_event as the Value field.

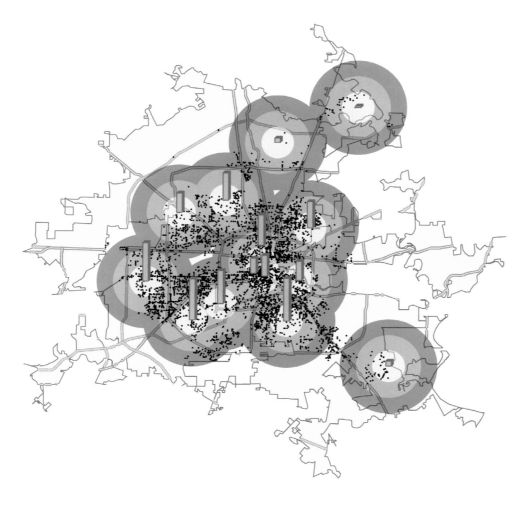

Q19 *What does this method tell the police department about the distribution of crime?*

Q20 *Compare the two methods of analysis.*

Q21 *Write a spatial analysis of crimes and police stations.*

11. Save your map document as crime2.

12. Save the map document again as crime3.

Deliverable 2: A map of police stations and locations of crimes along with recommendations for new police substations sites.

STEP 4: Produce density maps of auto theft, burglary, and narcotics-related crimes

You will use the ArcGIS Spatial Analyst extension for the next part of the project. ArcGIS Spatial Analyst allows you to calculate a continuous distribution of a particular type of crime from a set of input points. This density "surface" provides data throughout your area of interest and gives you a better indication of the distribution of crime in that area. Density maps are frequently used in crime analysis to show where crimes are concentrated and to aid the search for patterns.

MAKING
SPATIAL
DECISIONS
USING GIS

Law
enforcement
decisions

1. Open map document crime3 and be sure that the ArcGIS Spatial Analyst extension is turned on.

2. Remove police_stations, crime buffers, crime bar graphs, and highways.

To calculate a density map, the computer needs to have a numerical value for each event. (Text fields won't work.) Remember that you added an Event field and a 1 for each crime in step 3. You are now ready to create a density map. The police department is particularly interested in aggravated assault, burglary, and narcotics-related crime.

3. Select by Attribute all the burglaries. Do not include "Burglary of a Motor Vehicle."

Q22 How many burglaries have been reported?

4. Construct a kernel density map with crime as the input feature and Event as the population field. The output raster should be named burglary, the output cell size is 200, the search radius is set to 5,000, and the Area units are SQUARE_KILOMETERS.

5. Clear the selected features in crime and turn crime off.

Density maps can be hard to interpret in the initial symbology provided by ArcMap. Statistically, the map says that in the darkest green area there are 3.8 burglaries per square kilometer. It is easier to interpret the map if you classify the crime density using natural breaks.

6. For this burglary density map and the other density maps, classify the data using natural breaks with five classes labeled as follows:
    1. Low
    2. Low Medium
    3. Medium
    4. High Medium
    5. High

7. Save the Layer File as crime. A layer file cannot be stored in a geodatabase. Put the layer file in the \Houston_results folder.

8. Create contours with burglary being the input raster, con_burglary being the output polyline feature and Contour intervals of 1. Contours are lines that connect points of equal value, which in this case shows concentration of burglaries.

9. Add World Imagery from ArcGIS Online and examine the areas with high density of burglaries. Do these areas have any features in common? (Hint: Look at the kinds of buildings, amount of open space, density of housing, etc.)

Q23 **What is the spatial distribution of burglaries in Houston?**

10. Name the data frame burglary.

11. Insert another data frame and name it narcotics. Remember to set the coordinate system for the data frame properties. Repeat the above procedure for narcotics-related crimes. Import the crime layer file to use the same symbology for these crimes. Create contours and name them con_narcotics. Use a contour interval of 1.

Q24 **What is the spatial distribution of drug crimes in Houston?**

12. Repeat the above procedure for aggravated assault. Use a contour interval of 1.

Q25 **Describe the spatial distribution of aggravated assault crimes in the Houston area.**

13. Produce a presentation map showing aggravated assault, burglary, and narcotics-related crimes.

14. Add the census block groups (blkgrp) to each of the crime data frames.

15. Symbolize the census block by quantities using population density (POP00_SQMI) as the value field.

16. Move the crime density layers above the population density layers.

17. Turn on the Effects toolbar and use the Swipe tool to compare the crime density map with the population density. The Swipe tool allows you to reveal what's underneath a particular layer. Make sure the top layer is shown in the Effects toolbar.

18. Save the map document as crime3.

19. Save map document again as crime4.

Deliverable 3: Density maps of aggravated assault, burglary, and narcotics-related crimes.

STEP 5: Produce time crime maps

1. Open the map document crime4.

2. Remove the Narcotics and Aggravated Assault data frames.

3. Rename the Burglary data frame Time and remove Highways, Burglary, blkgrp, and contours.

4. Add crime.

When you open the crime attribute table, you will find that Offense_Ti (time) is given on a 24-hour clock. The table below helps you convert time to a 12-hour clock.

12-hour Clock	24-hour Clock	12-hour Clock	24-hour Clock
12:00 midnight	0000 hrs	12:00 noon	1200 hrs
1:00 AM	0100 hrs	1:00 PM	1300 hrs
2:00 AM	0200 hrs	2:00 PM	1400 hrs
3:00 AM	0300 hrs	3:00 PM	1500 hrs
4:00 AM	0400 hrs	4:00 PM	1600 hrs
5:00 AM	0500 hrs	5:00 PM	1700 hrs
6:00 AM	0600 hrs	6:00 PM	1800 hrs
7:00 AM	0700 hrs	7:00 PM	1900 hrs
8:00 AM	0800 hrs	8:00 PM	2000 hrs
9:00 AM	0900 hrs	9:00 PM	2100 hrs
10:00 AM	1000 hrs	10:00 PM	2200 hrs
11:00 AM	1100 hrs	11:00 PM	2300 hrs
		12:00 midnight	2400 hrs

5. Open the attribute table and study the Offense_Ti field. Sort the records in ascending order for Offense_Ti.

This analysis will be easier if the time is expressed in hours and not hours and minutes. If you could truncate the last two digits, you would have just the hour in which the crime occurred.

6. Add field (Data Management) to crime. Crime is the Input Table, the Field Name is hours, and the Field Type is SHORT.

7. Right-click Hours and open the Field Calculator. Enter the following expression in the box in the Field Calculator window to retain only the first two digits of the time:

Left([Offense_Time],2)

8. Sort the Hours field in ascending order. It now starts at midnight, and the last hour is 23.

MAKING
SPATIAL
DECISIONS
USING GIS

*Law
enforcement
decisions*

Q26 ***What is the total number of crimes?***

Q27 ***Calculate crime for various time periods by completing this table on your worksheet:***
- Hours >= 0 AND Hours < 6 Night
- Hours >= 6 AND Hours < 12 Morning
- Hours >= 12 AND Hours < 18 Afternoon
- Hours >= 18 AND Hours <=23 Evening

Time	Name	Zulu Time (24-hour clock)	Crimes	Percent
Midnight–6:00 AM	night	00–06		
6:00 AM–noon	morning	06–12		
Noon–6:00 PM	afternoon	12–18		
6:00 PM–midnight	evening	18–24		

9. Clear Selected Features and Summarize the field heading hours. Name the dbf file totalcrime_hrs and save in \houston_projects.gdb. Add the totalcrime_hrs table to the table of contents.

10. Create a graph from a table representing crime for a 24-hour period (totalcrime_hrs). The graph should have the following parameters:
- Vertical Bar for the Graph type.
- Totalcrime_hrs for the Layer/Table.
- For the Value field select Count_hours.
- Label the x field by Hours.
- Label the y field Crimes.
- Give the graph an appropriate title.

MAKING
SPATIAL
DECISIONS
USING GIS

1

*Law
enforcement
decisions*

11. Because this is a temporal dataset, the time properties can be enabled to visualize time using a time slider. In this instance, time should be enabled on both the crime layer and the totalcrime_hrs graph.

   a. Increase the size of the crime dots to 10.

   b. Go to Properties and Enable Time Properties on Crime using HOURS as the Time field. Choose "Each feature has a single time field."

   c. Choose HOURS as the Time field.

   d. Because 0001 to 0023 is the hour format, choose YYYY as the field format and select 1 year as the Time Step Interval.

   e. The Calculate the Layer Time Extent value should be 0001 to 0023.

   f. On the toolbar, click the Open Time Slider window.

   g. Click the Time Slider options and choose 1 year for the Time Step Interval.

   h. Select 2010 (yyyy) for the Display date format.

   i. Slow down the speed under the Playback tab.

   j. Hit the play arrow and watch the time simulation of crime.

   k. Enable the Time Properties on totalcrime_hrs following the steps listed above.

   l. In addition, do the following:

      i. Check Display data cumulatively.

      ii. Under Advanced Properties on the Graph, select Axis, Bottom Axes and set the Maximum Value to be 24.

   m. Play the time simulation.

12. Save the map document as crime4.

13. Save map document again as crime5.

Q28 *What time of day is crime the lowest?*

Q29 *What time of day is crime the highest?*

Q30 *How do the patterns of crime change throughout the 24-hour period?*

14. Open crime5, delete the graph, and remove totalcrime_hrs.

15. Create a series of auto theft density maps. Using the procedure in step 4, produce auto theft density maps for six-hour periods (midnight–6:00 AM, 6:00 AM–noon, noon–6:00 PM, 6:00 pm–midnight). Save each density map and name them night, morning, afternoon, and evening, respectively. (Hint: Select Auto Theft and use the "AND" operator to select the appropriate time. For example: "Offense"='Auto Theft' AND "Hours">=6 AND "Hours"<12.)

16. Create maps using Kernel Density with Event as the Population field and 5000 as the Search radius. Pick the density map that has the highest auto theft (5.45) and use that map to define the symbology.
    a. Classify using Equal Interval and 5 Classes.
    b. The classes could be labeled Low, Low Medium, Medium, High Medium, and High.
    c. Format the labels to have only two decimals.
    d. Save this as a layer file and name it auto_theft .

17. Import this layer file into each of the other density layers to use the same symbology for each layer.

18. Use the Effects tool to compare the layers.

19. Save the map document as crime5.

Deliverable 4: A graph exploring the time of day crimes are committed.

Once your analysis is complete, you still need to develop a solution to the original problem and present your results in a compelling way to the police officials in this particular situation. The presentation of your various data displays must explain what they show and how they help solve the problem.

Presentation

There are many ways, ranging from simple to advanced, that you can use to prepare a presentation. Whatever method of presentation you choose has to include a report documenting your analysis and addressing the needs of the Houston Police Department. Make specific recommendations on where to build new substations and clearly show concentrations of different crimes. As you work through this book, try to use different types of presentation media. Pick the presentation medium that best fits your audience. Remember that the police officials probably lack your in-depth knowledge of GIS, so you will need to communicate your results in a way they will be able to understand and use.

Listed below are various presentation formats:
- Create a text document with inserted maps.
- Show your findings in a digital slide presentation.
- Use ArcGIS Explorer Desktop, which is a free, downloadable GIS viewer that provides an easy way to explore, visualize, share, and present geographic information. This software can be downloaded at:

    http://www.esri.com/software/arcgis/explorer/download.html

MAKING
SPATIAL
DECISIONS
USING GIS

Law
enforcement
decisions

MAKING
SPATIAL
DECISIONS
USING GIS

1

Law
enforcement
decisions

- Embed interactive maps in your text documents. These maps can also be shared with others. The links below provide information about this tool:

  http://www.esri.com/software/mapping_for_everyone/index.html

  http://help.arcgis.com/en/webapi/javascript/arcgis/index.html
- Create layer packages or map packages and share your maps with your classmates.
- Use ArcGIS Explorer Online to produce an interactive online geospatial presentation that can be shared: http://www.arcgis.com/home/

Extending the project

Your instructor may choose for you to complete this optional exercise to evaluate drug-free zones.

In the state of Texas, there are drug-free zones within 1,000 feet of schools. Anyone arrested for dealing drugs within these zones faces extra-stiff mandatory penalties. To document the effectiveness of these laws, calculate the number of crimes within 1,000 feet of the schools in Houston.

Q31 **How many schools do not have any crime within the 1,000-foot zone?**

Q32 **What school has the highest number of crimes nearby?**

Q33 **Select that school and see which crimes were committed.**

Q34 **Is the drug-free zone effective?**

Q35 **How could you design a study to find out what is going on just outside the zone?**

References

Chainey, S. and J. Ratcliffe (2005). *GIS and Crime Mapping.* John Wiley and Son, Chichester, UK.

Hirschfield, A. and S. Bowers (2001). *Mapping and analyzing crime data: Lessons from research and practice.* Taylor and Francis, London.

MAKING
SPATIAL
DECISIONS
USING GIS

2

*Law
enforcement
decisions*

PROJECT 2
Logging Lincoln's police activity

Scenario

In this exercise you have been hired as a consultant to the Lincoln, Nebraska, Police Department to help assess the allocation of police resources and the geographic distribution of crimes.

Problem

The Lincoln Police Department knows the value of using GIS for crime analysis and wants to investigate the following:

- Crime patterns in proximity to police stations to determine if current patrols are effective or if adjustments are necessary, such as reorganizing beats and building new substations.
- Patterns of crime by days of the week.
- Patterns of assault, burglary, and shoplifting.
- Patterns of when crimes are committed.

85

Reminder: It helps to divide this large problem into a set of smaller tasks, such as the following:

- Identify the geographic study area.
- Determine the sequence of steps in your study.
- Identify the decisions to be made.
- Develop the information required to make decisions.
- Identify stakeholders for this issue.

MAKING
SPATIAL
DECISIONS
USING GIS

2

Law
enforcement
decisions

The questions in this project are both quantitative and qualitative. They identify key points that should be addressed in your analysis and presentation.

Deliverables

We recommend the following deliverables for this exercise:

1. A basemap of the Lincoln area showing police stations, crimes, roads, and census blocks in August 2006. Include a graph of total crimes by days of the week.
2. A map of police stations and locations of crimes along with a recommendation of where to build new substations.
3. Density maps of assault, burglary, and shoplifting. Maps should have contour lines highlighting high-crime areas along with aerial or satellite imagery.
4. A graph exploring crime throughout a 24-hour period. Daily crime should be viewed temporally.
5. Density maps of aggravated assault, burglary, and narcotics-related crimes. Maps should have contour lines highlighting high crime areas along with aerial or satellite imagery.
6. Graphs and maps to explore the time of day crimes are committed. Daily crime should be viewed temporally.

Examine the data

The data for this project are stored in the **\Project2_Lincoln\Lincoln_data** folder.

Reminder: View the item description to investigate the data. The table below helps you organize this information.

Q1 Investigate the metadata and complete the following table on your worksheet.

Layer	Publication Information: Who Created the Data?	Time Period Data Are Relevant	Spatial Horizontal Coordinate System	Data Type	Resolution for Rasters	Attribute Values
aug_06	Lincoln Police Department	2006	N/A	database	N/A	
usa_streets	Tele Atlas North America, Inc., Esri		Geographic		N/A	Street Data

Organize and document your work

Be sure to refer to the project 1, Houston exercise and your process summary.

1. Set up the proper directory structure.
2. Create a process summary.
3. Document the map.
4. Set the environments:
   a. Set the Data Frame Properties Coordinate System to UTM NAD 1983 Zone 14N.
   b. Set the working directory.
   c. Set the scratch directory.
   d. Set the Output Coordinate System to Same as Display.
   e. Set the Extent to Same As Layer blkgrp or Lincoln.
   f. Set the Output Cell Size to 200.
   g. Set the Mask to blkgrp or Lincoln.

Analysis

An important first step in GIS analysis is to develop a basemap of your study area. Complete deliverable 1 and answer the questions below to orient yourself to the study area.

Deliverable 1: A basemap of the Lincoln area showing police stations, crimes, roads, and census blocks in August 2006. Include a graph of total crimes by days of the week.

Reminder: Be sure to create an address locator before you geocode the police stations. The crimes have already been geocoded.

Q2 Were all the police stations matched when they were geocoded?

MAKING
SPATIAL
DECISIONS
USING GIS

2

*Law
enforcement
decisions*

Q3   How many crimes have the day of the week recorded? Not recorded?

Q4   What percent did not record the day of the week?

***Q5   Looking at the graph of crime by days of the week, what day(s) would
require more of a police presence?***

Deliverable 2: A map of police stations and locations of crimes along with a recommendation of
where to build new police substations.

Reminder: Make sure you complete the following:
- Set the buffer unit to miles and the distances to 1, 2, and 3.
- Add a field named Event and enter a **1** for each crime.
- Spatially join the crimes to the police station.
- Use all the crimes for this deliverable, not just the ones with day recorded.

Q6   How many total crimes are there?

Q7   Complete this table on your worksheet.

Distance in Miles	Crimes	Percent
1		
2		
3		
More than 3		

***Q8   What does this method tell the police department about the distribution of
crime?***

MAKING

SPATIAL

DECISIONS

USING GIS

2

Law

enforcement

decisions

Q9   *Complete this table on your worksheet.*

Police Station	Number of Crimes	Percent
'F' Street Community Center		
Auld Recreation center		
Bess Dobson Walt Library		
Bryan/LGH		
Headquarters		
Highlands		
LMEF Inc.		
LPD Center Team Station		
Northeast Team Station		
Team Station		
Union College		

Q10   *What does this method tell the police department about the distribution of crime?*

Q11   *Compare the two methods of analysis.*

Q12   *Write a spatial analysis of crimes and police stations.*

Deliverable 3: Density maps of assault, burglary, and shoplifting. Maps should have contour lines highlighting high crime areas along with aerial or satellite imagery.

Reminder: Be sure to check these settings.
- Set the coordinate system for each data frame.
- Use Event as the Population field.
- Use a radius of 1,000.
- Use all crimes for this deliverable, not just the ones with day recorded.

Study the document UCR_codes.txt in the documents folder. In the attribute table, TYPE_CODE represents the UCR codes.

Q13   *What are the code numbers for assault?*

Q14   *What are the code numbers for burglary?*

MAKING
SPATIAL
DECISIONS
USING GIS

2

*Law
enforcement
decisions*

Q15 ***What are the code numbers for shoplifting?***

Q16 ***How many assaults occurred?***

Q17 ***How many burglaries?***

Q18 ***How many instances of shoplifting?***

Reminder: Classify the data using natural breaks with five classes. For all three crimes, create a single standard layer file of **Low**, **Low Medium**, **Medium**, **High Medium**, and **High**.

Q19 ***Describe the spatial distribution of assault in Lincoln.***

Q20 ***Describe the spatial distribution of burglary in Lincoln.***

Q21 ***Describe the spatial distribution of shoplifting in Lincoln.***

Deliverable 4: A graph that displays crime throughout a 24-hour period. Daily crime should be viewed temporally.

Reminder: Use the Field Calculator to change the Zulu time to hours. Use the TIME_FROM for your Time field and be sure to select only the records that have recorded times. You must not select any record that is not a standard Zulu time. In this database there are records that do not have a recorded time.

There are also records that have unk or UNK, and there are incorrect records of 7130, 8-12, and AM. DO NOT USE THESE RECORDS. Make sure your selection contains only records with times 0000 to 2400. The easiest way to do this is to select all the files that have a recorded time and export them to a new data file.

Q22 ***Complete this table on your worksheet.***

Time	Name	Zulu Time (24-hour clock)	Crimes	Percent
Midnight–6:00 AM	night	00–06		
6:00 AM–noon	morning	06–12		
Noon–6:00 PM	afternoon	12–18		
6:00 PM–midnight	evening	18–24		

MAKING
SPATIAL
DECISIONS
USING GIS

2

*Law
enforcement
decisions*

Q23 *How do the patterns of crime change throughout the 24-hour period?*

Q24 *What is the total number of crimes with times?*

Q25 *What time of day is crime the lowest?*

Q26 *What time of day is crime the highest?*

Reminder: Check your time settings.
- Hours >= 0 AND Hours < 6 (night)
- Hours >= 6 AND Hours < 12 (morning)
- Hours >= 12 AND Hours < 18 (afternoon)
- Hours >= 18 AND Hours <=24 (evening)

Summarize field heading Hours. Create a graph from summarized.dbf using count as the Value field. Enable time on time by using Hours as Time field. Enable time on summarized.dbf cumulatively with Hours being the Time field.

Presentation

Keep in mind the interests and expertise of your audience as you prepare your presentation. Develop a solution to the original problem and present your results in a compelling way.

Refer to the list of presentation format options in project 1.

PROJECT 3
On your own

You have worked through a guided activity on the variation of crime in a large urban area and repeated the analysis with data from a smaller city. In this section, you will reinforce the skills you have developed by researching and analyzing a similar scenario entirely on your own. First you must identify your study area and acquire the data for your analysis. Links to three jurisdictions with crime data posted online are listed below. However, if there is another, more local area that has significant interest for you, locate, download, and work with that data. If you plan to work with your local area, be sensitive to the fact that some of the reports contain individual names which should be removed before you present any results.

- Sacramento, California: http://www.cityofsacramento.org/gis/
- Houston, Texas: http://www.houstontx.gov/police/stats.htm
- Fairfax County, Virginia: http://www.fairfaxcounty.gov/police/

Refer to your process summary and the preceding module projects if you need help. Here are some basic steps to help you organize your work.

Research

Research the problem and answer the following questions:

1. What is the area of study?
2. What data are available?

Obtain the data

Do you have access to baseline data? The Esri Data & Maps Media Kit provides many of the layers of data that are needed for project work. Be sure to pay particular attention to the source of data and get the latest version. Obtaining crime data can be challenging. Much of the data are summarized or displayed in PDF format. If you are interested in doing a local crime study, ask your local police department for help. Tell your police department that you are not interested in individual names or addresses. Data by block are generally more than sufficient for your analysis.

If you do not have access to the Esri Data & Maps Media Kit, you can obtain data from the following sources:

- Census 2000/TIGER/Line data http://www.esri.com/tiger
- Geospatial One Stop http://gos2.geodata.gov/wps/portal/gos
- The National Atlas http://www.nationalatlas.gov

MAKING
SPATIAL
DECISIONS
USING GIS

3

Law
enforcement
decisions

Workflow

After researching the problem and obtaining the data, you should do the following:

1. Write a brief scenario.
2. State the problem.
3. Define the deliverables.
4. Examine the data by using ArcCatalog.
5. Set the directory structure, start your process summary, and document the map.
6. Decide what you need for the data frame coordinate system and the environments.
   a. What is the best projection for your work?
   b. Do you need to set a cell size or mask?
7. Start your analysis.
8. Prepare your presentation and deliverables.
9. Always remember to document your work in a process summary.

Presentation

Refer to the list of presentation format options in project 1.

HURRICANE DAMAGE
DECISIONS

Introduction

In 2005, Hurricanes Katrina, Rita, and Wilma destroyed homes, businesses, infrastructure, and natural resources along the Gulf of Mexico and Atlantic coasts. In the aftermath of the storms, federal, state, and local governments; service agencies; and the private sector responded by helping to rebuild the hurricane-ravaged areas and restore the local economies. GIS helped responders assess damage, monitor the weather, coordinate relief efforts, and track health hazards—among many other critical tasks—by providing relevant and readily available data, maps, and images. In this module, you will access some of the same data that guided critical decisions, such as funding and safety measures, in the wake of Hurricanes Katrina and Wilma. You will map elevations and bathymetry, animate each hurricane's path and changes in pressure, analyze flooded areas and storm surges, and pinpoint vulnerable infrastructure. In the real world, this process saves time, money, and most importantly, lives.

Scenarios in this module

- Coastal flooding from Hurricane Katrina
- Hurricane Wilma storm surge
- On your own

GIS software required

- ArcGIS Desktop 10 (ArcEditor)
- ArcGIS Spatial Analyst

Student worksheets

The student worksheet files can be found on the Data and Resources DVD.

Project 1: Katrina student sheet
- File name: Katrina_student_worksheet.doc
- Location: Project1_Katrina\Documents

Resource material: NLCD classification in a Microsoft Excel spreadsheet
- File name: NLCD_classification.xls
- Location: Project1_Katrina\Documents

Project 2: Wilma student sheet
- File name: Wilma_student_worksheet
- Location: Project2_Wilma\Documents

PROJECT 1
Coastal flooding from Hurricane Katrina

The Gulf Coast contains one of the nation's most fragile coastal ecosystems. The barrier islands, coastal wetlands, and forest wetlands each play a critical role in the environment. The barrier islands prevent storm surge and saltwater intrusion. Coastal wetlands provide habitat for mammals and waterfowl while also serving as a nursery for the Gulf's fishing industry. Forest wetlands provide a renewable resource for the paper and timber industry.

Scenario

On Monday, August 29, 2005, Hurricane Katrina hit the Gulf Coast, devastating the wetlands and barrier islands. Three counties in Mississippi were particularly hard hit. To assess and remediate the damage caused by the 15-foot storm surge that hit the coast, your team of GIS specialists must prepare maps and calculate damage. The information will help federal officials decide how to allocate redevelopment resources. In the aftermath of the hurricanes of 2005, considerable thought has been given to better preparing for such disasters. Professor Raymond J. Burby (2006) concluded in a paper

on governmental planning in hazardous areas that disaster losses can be minimized if local governments address hazard mitigation in comprehensive plans.

Problem

Federal officials need to decide where to allocate disaster aid to the Mississippi counties most affected by Hurricane Katrina. You will assess the total acreage of different types of land cover that were under water as a result of the Katrina storm surge. The damage reflects the unique hydrography of the Gulf Coast, so you must map that as well. You also will map damage to the infrastructure and health-care centers so that restoration efforts can focus on areas with the greatest need. Your team must prioritize locations for disaster aid, justified by the data provided and the maps, graphs, and tables you produce. As with many of the complex problems that confront us, there is not a single "right" answer. However, some answers are better supported and justified than others.

When applying GIS to a problem, it is critical that you have a very clear understanding of the situation. We find it helpful to answer these four questions that test your understanding and divide the problem into smaller problems that are easier to solve. Record your answers on the worksheet provided.

Q1   What geographic area are you studying?

Q2   What decisions do you need to make?

Q3   What information would help you make the decisions?

Q4   Who are the key stakeholders for this issue?
(This step is important. You need to know the audience for your analysis to help decide how to present your results.)

Deliverables

After identifying the problem, you need to envision the kinds of data displays (maps, graphs, and tables) that will provide the solution. We recommend the following deliverables for this exercise:

1. A map showing elevation/bathymetry of Mississippi coastal counties with places and types of water, barrier islands, and hydrography.
2. A time series map showing Katrina's path with graphs of wind speed and air pressure.
3. A map of flooded land of the Mississippi coast after Hurricane Katrina. The map should include a bar graph showing percentage of total flooded land by land-cover type.
4. A map showing infrastructure and health facilities at risk from the storm surge.

MAKING
SPATIAL
DECISIONS
USING GIS

1

Hurricane
damage
decisions

5. A table showing various land types that were flooded, measured in acres and square miles.

6. False color composite images and derived land cover created using Landsat Thematic Imagery.

Tips and tools

MAKING

SPATIAL

DECISIONS

USING GIS

Hurricane

damage

decisions

Topical instructions are given in the following exercises. If more detailed instructions are needed, ArcGIS Desktop 10 provides these options:

1. Use the Help file to ask a question or look up a keyword, such as a tab, menu option, or function. If you are online, it is better to use the Web-based help option by accessing the ArcGIS Resource Center for an up-to-date version of the help system included with the software.

2. The tools can be accessed by using the traditional ArcToolbox or you can use the Search For Tools option found under the Geoprocessing menu.

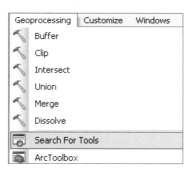

When you search for the tools, an explanation and a link to the tool appear.

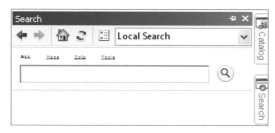

The questions in this project are both quantitative and qualitative. They identify key points that should be addressed in your analysis and presentation.

Examine the data

The next step in your workflow is to identify, collect, and examine the data for the Katrina analysis. Here, we have identified and collected the data layers you will need. Explore the data to better understand both the raster and vector feature classes in this exercise. In order to explore the data, you need to access the metadata associated with each feature class. Thoroughly investigate the data layers to understand how they will help you address the problem. The spatial coordinate system, the resolution of the data, and the attribute data are all important pieces of information about a feature class.

MAKING
SPATIAL
DECISIONS
USING GIS

1

*Hurricane
damage
decisions*

1. Open ArcMap. (For these exercises, the Getting Started dialog box is not needed. Select the "Do not show this dialog in the future" option.)

2. Add Data by connecting to the folder \Project1_Katrina\Katrina_data. **The data folder contains the Katrina geodatabase, which holds twelve feature classes and two rasters. Add the features airports, churches, counties, and hospitals and the rasters elev and landcover.**

There are different metadata styles that control how you view an item's description. The metadata data style "ISO 19139 Metadata Implementation Specification" supports formal metadata and allows the complete metadata to appear in the Item Description window. This has to be set in the stand-alone ArcCatalog software.

3. Open ArcCatalog and in the Customize menu choose ArcCatalog Options. Select the Metadata tab and select "ISO 19139 Metadata Implementation Specification" from the Item Description menu.

4. Close ArcCatalog.

Q5 **View the item descriptions for these features and complete the following table on your worksheet.**

Layer	Data Type	Publication Information: Who Created the Data?	Time Period Data Are Relevant	Spatial Horizontal Coordinate System	Attribute Values	Resolution for Rasters
airports	Vector				N/A	N/A
elev				Geographic: GCS_North_ American_1983	Elevation is given in meters	1 arc second, 0.000278 decimal degrees, or 30 meters
landcover	Raster		2001			

In some cases, the values of the attributes can be difficult to understand without supplementary information. This is particularly true for the landcover raster. The U.S. Geological Survey National

99

Land Cover Class definitions are given in the file NLCD_classification.xls (Microsoft Excel format) in the **student_resources** folder on the DVD.

5. Close ArcMap.

Now that you have explored the available data, you are almost ready to begin your analysis. First you need to start a process summary, document your project, and set the project environments.

MAKING
SPATIAL
DECISIONS
USING GIS

Hurricane
damage
decisions

Organize and document your work

The following preliminary steps are essential to a successful GIS analysis.

Examine the directory structure

The next phase in a GIS project is to keep track of the data and your calculations. You will work with different files and must keep them organized so you can easily find them. The best way to do this is to have a folder for your project that contains a data folder. For this project, the folder named **\Project1_Katrina\Katrina_data** will be your project folder. Make sure that it is stored in a place where you have write access. Store your data inside the results folder. The results folder already contains an empty geodatabase named **\Katrina_results**. Save your map documents inside the **\Katrina_results** folder.

Create a process summary

The process summary is simply a list of the steps you used to do your analysis. We suggest a simple text document for your process summary. Keep adding to it as you do your work to avoid forgetting any steps. The list below shows an example of the first few entries in a process summary:

1. Explore the data.
2. Produce a map with elevation and bathymetry, counties, places, and islands.
3. Create time series of Hurricane Katrina tracks with graphs of wind speed and air pressure.
4. Change elevation from meters to feet.
5. Isolate the flooded land.

Document the map

1. Open ArcMap and save a new map document as katrina1. **Save it in the** \Katrina_results **folder.**

You need to add descriptive properties to every map document you produce. Use the same descriptive properties for every map document in the module or individualize the documentation from map to map. You can access the Map Document Properties from the File menu. The Document Properties dialog box allows you to add a title, summary, description, author, credits, tags, and hyperlink base. After writing descriptive properties, be sure to select the Pathnames check box, which makes ArcMap store relative paths to all of your data sources. Storing relative paths allows ArcMap to automatically find all relevant data if you move your project folder to a new location or computer.

Set the environments

In GIS analysis, you will often get data from several sources and these data may be in different coordinate systems and/or map projections. When using GIS to perform area calculations, you would like your result to be in familiar units, such as miles or kilometers. Data in an unprojected geographic coordinate system have units of decimal degrees, which are difficult to interpret. Thus, your calculations will be more meaningful if all the feature classes involved are in the same map projection. Fortunately, ArcMap can do much of this work for you if you set certain environment variables and data frame properties. In this section, you will learn how to change these settings. To display your data correctly, you will need to set the coordinate system for the data frame. When you add data with a defined coordinate system, ArcMap will automatically set the data frame's projection to match the data. If you add subsequent layers that have a coordinate system different from the data frame, they are automatically projected on-the-fly to the data frame's coordinate system.

1. From the View menu, choose Data Frame Properties. Click the Coordinate System tab. Click Import. Navigate to your data folder, select landcover from the Katrina geodatabase, and click Add.

2. From the Geoprocessing menu, choose Environment. Remember that the environment settings apply to all the functions within the model. The analysis environment includes the workspace where the results will be placed, and the extent, cell size, and coordinate system for the results.

3. Expand workspace. By default, inputs and outputs are placed in your current workspace, but you can redirect the output to another workspace such as your results folder. Set the Current Workspace as \Project1_Katrina\Katrina_data\Katrina.gdb. Set the Scratch Workspace as \Project1 _Katrina\Katrina_results\Katrina_results.gdb.

4. For Output Coordinate System, select Same as Display.

MAKING
SPATIAL
DECISIONS
USING GIS

1

Hurricane
damage
decisions

5. For Processing Extent, select Same as Layer counties.

6. Expand the Raster Analysis Settings and set the Cell size to 30.

You also need to limit your analysis to the three Mississippi coastal counties most affected by the storm. This is accomplished by using an analysis mask. The mask identifies those locations within the analysis extent that will be included when using a tool.

7. Set the Mask to Counties from the Katrina geodatabase.

8. Click OK and save the project again as katrina1.

Analysis

Once you have examined the data, completed map documentation, and set the environments, you are ready to begin the analysis and to prepare the displays to address the flooding problem. A good place to start any GIS analysis is to produce a basemap to better understand the distribution of features in the geographic area you're studying. First, you will prepare a basemap of the Mississippi coast showing the elevation of land and ocean floor.

STEP 1: Display the Mississippi coast's elevation, bathymetry, and hydrography

1. Add the Elevation raster (elev) to the map document. The elevation feature class shows elevation above sea level for land and bathymetry for the ocean floor.

Q6   What are the highest and lowest elevations?

2. Symbolize elev with an appropriate color ramp.

3. Add counties and make them hollow or transparent.

4. Label the counties. Change the Label Field from STATE to County in the Properties/Label Tab for counties.

Q7   What are the names of the counties in order from west to east?

5. Add places. Choose an appropriate symbol.

6. Add islands and label them.

7. Add water and symbolize it with Unique Values using the FTYPE field. Make the Swamp/Marsh type a distinct color or symbol.

8. Add rivers and make them an appropriate color.

9. Save the map document as katrina1.

10. Save the map document again as time. (When you save the map document again as time, the software correctly saves the documentation, the data frame projection, and the environment settings. This saves you from redoing these variables for other deliverables.)

Deliverable 1: A map showing elevation/bathymetry of Mississippi coastal counties with places, types of water, islands, and hydrography.

Q8 ***What does this map show you about the Gulf Coast?***

Q9 ***Describe the spatial distribution of features.***

Q10 ***How might these be affected by a large storm surge?***

STEP 2: Create a time series map showing Katrina's path with graphs of wind speed and air pressure.

1. Open the map document time.

2. Remove places, rivers, water, islands, and elevation. Remove the county labels.

3. Add states and track_katrina from the Katrina.gdb. Zoom to track_katrina. Rename track_katrina as katrina.

4. Symbolize Katrina by unique values using CAT as the Value field. The list below gives the storm categories. Save the symbolization as a layer file called Storms.

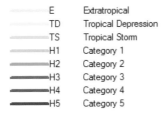

	E	Extratropical
	TD	Tropical Depression
	TS	Tropical Storm
	H1	Category 1
	H2	Category 2
	H3	Category 3
	H4	Category 4
	H5	Category 5

MAKING
SPATIAL
DECISIONS
USING GIS

1

Hurricane
damage
decisions

MAKING
SPATIAL
DECISIONS
USING GIS

1

Hurricane
damage
decisions

5. Open the attribute table for Katrina and investigate the time and time2 fields. Time shows the year, month, and day. Time2 shows the year, month, day, and hour. Time properties of your temporal data can be set using this information. Close the attribute table.

6. Set the time properties for Katrina under the Time tab in the Properties window. Do the following:
   - Check Enable time on this layer.
   - Enable each feature with a single time field.
   - The time field should be time2.
   - The Field format should be YYYYMMDDhhmmss.
   - The Time Step Interval should be 1 Hours.
   - Calculate Layer Time Extent: The Layer Time Extent value should be 8/23/2005 6 PM to 8/31/2005 12:00 AM.
   - Check the Display data cumulatively box and click OK.
   - The Time Slider window provides controls that allow you to visualize temporal data. The Time Slider window can be invoked by clicking the Open Time Slider Window button in the Tools toolbar.

Features of the Time Slider window

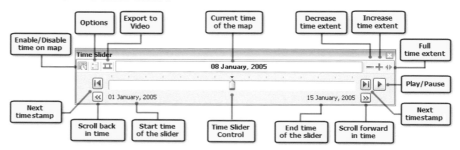

   - Click Options in the Time Slider and under the Playback tab, decrease the speed.
   - Click the Play tab. The hurricane should traverse its track, moving from the beginning to the end with the color changing to show the current storm category.

7. Copy track_katrina and paste as a new layer. Name the layer Wind Speed.

8. The next step in our time analysis is to create a graph by using the Graph Wizard with wind speed as the Value field. The graph should have the following properties:
   - Graph type: Vertical Line
   - Layer/Table: Wind Speed
   - Y field: WIND_KTS
   - Title: WIND SPEED
   - Bottom axis: Time (in six-hour intervals)
   - Left axis: Wind Speed
   - Click the top menu bar of the graph and choose Advanced Properties. Select axis and bottom axis, and change the maximum number to 30. Change the maximum of the left axis to 160. This allows for a smooth animation of the graph.
   - Enable Time on the Wind Speed layer as you did above for Katrina.

9. Repeat steps 7 and 8 using Pressure as the Value field.
   - Under Advanced Properties on the graph, select axis and bottom axis, and change the maximum number to 30. Change the maximum on the left axis to 1010. This allows for a smooth animation of the graph.

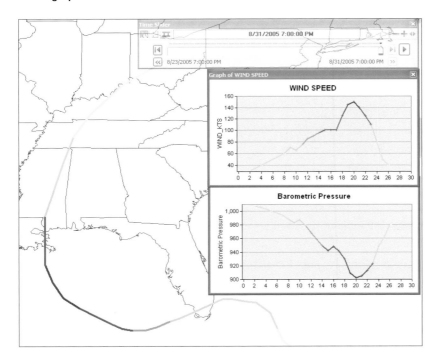

10. From the File menu, select Add ArcGIS Online, search for imagery, and add World Imagery.

11. Save the map document as time.

MAKING
SPATIAL
DECISIONS
USING GIS

1

Hurricane
damage
decisions

Deliverable 2: A time series map showing Katrina's path with graphs of wind speed and air pressure.

Q11 *Write an analysis listing all the variables and their relationships. For example, the relationship of wind speed to location on land versus water, or wind speed to depth of water, etc.*

With your first two deliverables complete, you are ready to begin the quantitative analysis of the problem.

Steps 3–6 will lead you through the identification of the types of land cover most affected by the storm surge.

STEP 3: Calculate flooded land

1. Open the map document katrina1 and save as katrina2.

2. Remove rivers and water.

3. Turn on the ArcGIS Spatial Analyst extension.

The elevation data is given in meters. The storm surge was given as 15 feet. To convert elev from meters to feet, multiply each elevation value in elev by 3.28 (3.28 feet = 1 meter).

4. Use the Times (Spatial Analyst) tool with elev as the input raster and 3.28 as the constant value 2. Save the output raster as elev_ft in your katrina_results geodatabase.

5. Remove elev from the Table of Contents.

6. The next step is to isolate the area flooded by the storm surge. Using the Less Than Equal tool, select elev_ft as the input raster and 15 as the constant value 2.

7. Save the output raster as flooded_land in your katrina_results geodatabase.

8. Remove elev_ft from the table of contents.

9. Save your map document as katrina2.

Anything that has a value of 1 fits the established flood criteria of less than or equal to 15 feet above sea level and thus represents the flooded land. Now that you have identified which land is flooded, you need to determine the type of land cover for the flooded land.

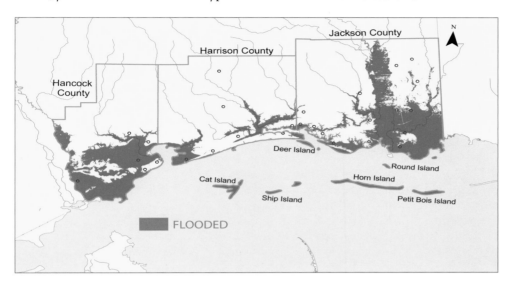

MAKING
SPATIAL
DECISIONS
USING GIS

1

*Hurricane
damage
decisions*

STEP 4: Reclassify national land cover

The sixteen separate land-cover classification categories in the landcover raster are too specific for your purposes, so you will group similar classifications to reduce the number of categories to six, which represents distinctly different kinds of land cover, as shown in the following table.

1. Add landcover and symbolize by unique values.

MAKING
SPATIAL
DECISIONS
USING GIS

1

Hurricane
damage
decisions

2. Using the Reclassify tool, choose value as the old value and type the reclassified value from the table below in the New Values column. Name the output raster reclandcover.

Original Values	Type of Land	Reclassed Values	Type of Land
11	Open Water	1	Water
21	Developed, Open Space	2	Developed
22	Developed, Low Intensity		
23	Developed, Medium Density		
24	Developed, High Intensity		
31	Barren Land	3	Barren
41	Deciduous Forest	4	Forest
42	Evergreen Forest		
43	Mixed Forest		
52	Scrub/Shrub	5	Agriculture
71	Grassland/Herbaceous		
81	Pasture/Hay		
82	Cultivated Crops		
90	Woody Wetlands	6	Wetlands
95	Emergent Herbaceous Wetland		
127	No Data	0	No Data

STEP 5: Isolate the flooded land cover

To identify the flooded land areas and classify it by land cover, you need to isolate the flooded areas. To do this, you must multiply the flooded_land by reclandcover.

1. Use the Times (Spatial Analyst) tool to multiply flooded_land by reclandcover.

2. Save the output raster as floodedlc in your katrina_results geodatabase.

3. Remove landcover, reclandcover and flooded_land from the Table of Contents.

4. Save your map document as katrina2.

STEP 6: Label the reclassified flooded land cover

1. Add a field (Data Management) to floodedlc. The input table is floodedlc, the Field Name is TYPE, the Field Type is text, and the Field Length is 15.

2. Open the floodedlc attribute table and start editing by turning on the Editor toolbar and enter the correct land cover for each type of land.

    Label the land as follows:

    0 = Not Flooded   4 = Forest
    1 = Water   5 = Agriculture
    2 = Developed   6 = Wetlands
    3 = Barren

3. Stop editing.

4. Display by Unique Values with the Value field TYPE and pick an appropriate color. Choose No Color for the value of 0. That land is not flooded.

5. Save as a layer file in order to save the symbology you just created.

6. Save the map document as katrina2.

MAKING
SPATIAL
DECISIONS
USING GIS

1

*Hurricane
damage
decisions*

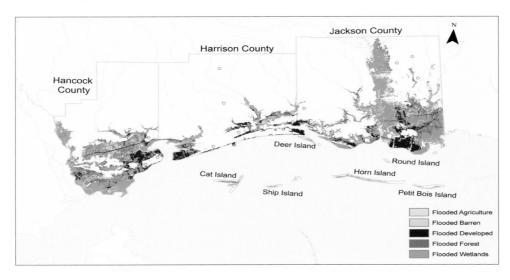

MAKING
SPATIAL
DECISIONS
USING GIS

1

*Hurricane
damage
decisions*

Q12 *Now that you have created the first part of deliverable 2, identify the extent of the flood damage, assess which areas were most affected, and summarize the damage to the counties and the barrier islands.*

The qualitative analysis above is a good start, but to make decisions about dealing with the damage from Hurricane Katrina requires a quantitative assessment of the acreage of each land type affected by the flood surge.

STEP 7: Determine percentage of flooded land type

In this step, you will create a graph that shows the percentage of each type of land that was flooded. The developed and residential land will cost the most to rebuild now, but the wetland destruction may have a longer lasting effect on the recovery of the economy and the environment. To determine percentages, you will use the count field in the attribute table of floodedlc. This field tells you the number of pixels of each land type and you will use that to measure area. You do not want to include land that was not flooded.

1. Open the floodedlc attribute table and select all types of land except Not Flooded.

2. Right-click the Count column header and select Statistics.

Q13 *Record the sum, which represents the number of pixels flooded.*

3. Add a field (Data Management) called Percent. The Field Type should be float.

4. Right-click the Percent field header and select Field Calculator. Perform the following calculation: COUNT/sum*100, **where sum is the value from Q13 above**

Remember: You want to calculate the percentage of land that is flooded. Make sure only the flooded land is selected.

Ignore the warning about calculating outside the editing session. It is always a good idea to verify that your percentages add up to 100.

STEP 8: Graph the percentage of flooded land by land-cover type

1. Create a graph from a table by using the floodedlc attribute table. The graph should have the following properties:
    *   Vertical bar graph with Percent as the Value field.
    *   Select Percent and Ascending for the x field.
    *   Label the x field with TYPE.
    *   Show only the selected features/records on the graph. (Do not show Not Flooded.)
    *   Enter the title Percentage of Flooded Land.
    *   View the graph in 3D.
    *   Add the graph to the layout by right-clicking the top blue bar.

2. Create an appropriate layout.

3. Save the map document as katrina2.

4. Save the map document again as katrina3.

MAKING
SPATIAL
DECISIONS
USING GIS

Hurricane
damage
decisions

Deliverable 3: A map of flooded land of the Mississippi coast after Hurricane Katrina. The map should include a bar graph showing the percentage of total flooded land by land-cover type.

Q14 *What does the graph tell you about the greatest impact of the storm surge?*

Q15 *Describe the distribution of the flooded areas and how the flood might affect the long-term sustainability of the region.*

The storm surge affected more than just the wetlands. A variety of infrastructure elements affect a region's economy and people's well-being. In this next step, you will explore the spatial distribution of the infrastructure in this region.

STEP 9: Prepare a map showing infrastructure and health facility destruction

As in any disaster, health facilities and the infrastructure that enables people to access them are critical to evacuation and recovery.

1. Open the map document katrina3.

2. Remove both Flooded Land Cover Layers, places, and islands.

3. Add usa_streets and import the Streets.lyr file. Label the major interstate and U.S. highways.

4. Add railroads and symbolize appropriately.

5. Add hospitals and symbolize appropriately.

6. Add churches and symbolize appropriately.

7. Add flooded_land and pick appropriate colors for flooded and non-flooded land.

8. Save the map document as katrina3.

9. Save the map document again as katrina4.

Deliverable 4: A map showing infrastructure and health facilities at risk from the storm surge.

Q16 **Describe the distribution of infrastructure and health facilities and which elements have likely been damaged in the storm surge.**

Q17 **How might you prioritize the damaged elements that should be restored first?**

STEP 10: Calculate acreage

The final step in your analysis is to calculate the acreage of the flooded areas. This information will allow your team, other government agencies, and insurance firms to make damage assessments. When you set the data frame coordinate system to that of the landcover raster (Albers conic equal area), all of the cells have a spatial unit of meters.

1. Open the map document katrina4.

2. Remove all layers except counties.

3. Add floodedlc.

These are the facts that you know:
- Each cell (pixel) has a dimension of 30×30 meters.
- There are 4,046.68 square meters in an acre.
- There are 640 acres in a square mile.

Q18 *If you add a field called **ACRES** and use the Field Calculator, write the formula that you would use to calculate acres.*

Q19 *If you add a field called **SQMILES**, write the formula that you would use to calculate square miles.*

4. Add a field (Data Management). Name the field ACRES and for Field Type select float.

5. Use the field calculator to perform the following calculation:
   ACRES = COUNT*900/4046.68

6. Add a field (Data Management). Name the field SQMILES and for Field Type select float.

7. Use the Field Calculator to perform the following calculation:
   SQMILES = ACRES/640

8. From your earlier analysis, the land-cover types are in the following order (least to greatest area):
   - Barren
   - Water
   - Agriculture
   - Forest
   - Developed
   - Wetlands

Q20 *Complete this table on your worksheet.*

Type	Acres	Sq. Miles

9. Save the project showing acreage as katrina4.

10. Save the map document again as katrina5_sept.

MAKING
SPATIAL
DECISIONS
USING GIS

1

Hurricane
damage
decisions

Deliverable 5: A table showing various land types that were flooded, measured in acres and square miles.

STEP 11: Use Landsat imagery for analysis

MAKING
SPATIAL
DECISIONS
USING GIS

1

Hurricane
damage
decisions

The NASA Landsat program continually collects images of the earth using filters of different wavelengths. The images are collected in bands with intensity represented by shades of gray. Color imagery is produced by combining three single-band images into a single image, assigning one color per band (red, green, blue). A true color band combination (RGB, Band 3, Band 2, Band 1) produces a natural-looking image. A band combination that combines near-infrared Band 4 with visible Bands 3 and 2 (RGB, Band 4, Band 3, Band 2) results in more well-defined water boundaries and senses peak chlorophyll reflectance (Band 4) as red and densely populated urban areas in light blue. Band combinations of multispectral images are used for detecting different objects in the image such as bare soils, roads, buildings, and vegetation. Band combinations can also be processed by classifying certain features. In this part of the exercise you will use image processing to do the following:

- Create a True Color Composite: An image that appears as it would to the human eye (321-RGB).
- Create a False Color Composite: An image where infrared is assigned to the red band (432-RGB).
- Analyze band combinations by using Unsupervised Classification (Iso Cluster): Grouping individual pixels or spatially grouped sets of pixels representing a feature such as soil, water, asphalt, etc. Resulting image represents land-cover classes.

For further information see the following Web sites:
- NASA Zulu at https://zulu.ssc.nasa.gov/mrsid/tutorial/Landsat%20Tutorial-V1.html
- Remote sensing tutorial at http://rst.gsfc.nasa.gov/Front/overview.html

A. Create a True Color Composite (321-RGB)

1. Open the map document katrina5_sept and remove all layers except counties.

2. Using composite bands, add Band 3, Band 2, and Band 1 from \Katrina_data\Landsat5\Sept_2005.

3. Name the file Sept321 and save in \Katrina_results\Landsat5\Sept_2005.

Data courtesy of the U.S. Geological Survey.

MAKING
SPATIAL
DECISIONS
USING GIS

Hurricane
damage
decisions

B. Create a False Color Composite (432-RGB)

1. Using composite bands, add Band 4, Band 3, Band 2.

2. Name the file Sept432 and save in \Katrina_results\Landsat5\Sept_2005.

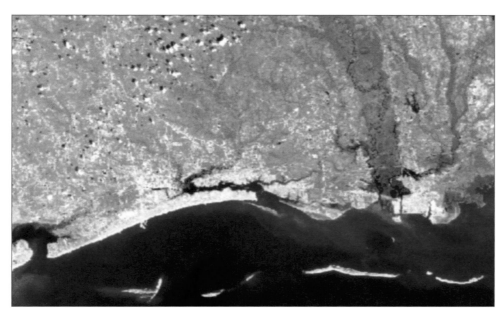

Data courtesy of the U.S. Geological Survey.

C. Analyze band combinations by using Unsupervised Classification (Iso Cluster)

1. Perform an Iso Cluster unsupervised classification with Sept432 as the input. Specify 5 as the Number of classes. Name the file unSept and save in \Katrina_results\Landsat\Sept_2005.

2. Using the Effects toolbar, swipe the images and compare how vegetation, wetlands, and urban areas are shown in both images. Bring in imagery from ArcGIS Online and create a legend identifying the different land classes. For example, the land classes could be water, wetlands, forest, urban/developed, and agriculture.

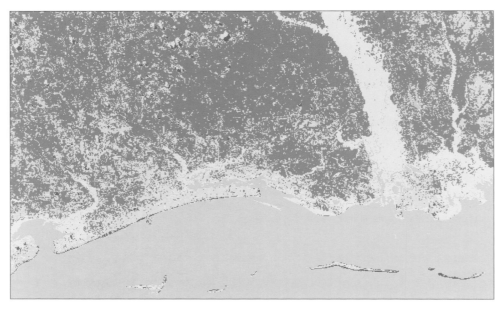

Data available from the U.S. Geological Survey.

Q21 *Write an analysis comparing how different land classes appear in the true and false color composite. Include in your analysis a discussion of the accuracy of the land-cover classification as you compared it to both true color, false color, and the aerial photography brought in from ArcGIS Online.*

Deliverable 6: False color composite images and derived land cover created using Landsat Thematic Imagery.

Once your analysis is complete, you still need to develop a solution to the original problem and present your results in a compelling way to the federal officials in this particular situation.

The presentation of your various data displays must explain what they show and how they help solve the problem.

Presentation

There are many ways, ranging from simple to advanced, that you can use to prepare a presentation. Whatever method of presentation you choose has to include a report documenting your analysis and addressing how to allocate redevelopment funds to human, infrastructure, or natural resources. You must explain the spatial patterns you see and describe the implications of your calculations and analysis for this problem. As you work through this book, try to use different types of presentation media. Pick the presentation medium that fits your audience. Remember that your audience probably lacks your in-depth knowledge of GIS, so you will need to communicate your results in a way they will be able to understand and use.

MAKING

SPATIAL

DECISIONS

USING GIS

1

Hurricane

damage

decisions

Listed below are various presentation formats:
- Create a text document with inserted maps.
- Show your findings in a digital slide presentation.
- Use ArcGIS Explorer Desktop, which is a free, downloadable GIS viewer that provides an easy way to explore, visualize, share, and present geographic information. This software can be downloaded at:

    http://www.esri.com/software/arcgis/explorer/download.html
- Embed interactive maps in your text documents. These maps can also be shared with others. The links below provide information about this tool:

    http://www.esri.com/software/mapping_for_everyone/index.html

    http://help.arcgis.com/en/webapi/javascript/arcgis/index.html
- Create layer packages or map packages and share your maps with your classmates.
- Use ArcGIS Explorer Online to produce an interactive online geospatial presentation that can be shared: http://www.arcgis.com/home/

Extending the project

Your instructor may have you complete these optional exercises: Add layers to ArcGlobe and analyze flooded areas by county.

Use the ArcGIS 3D Analyst extension and open ArcGlobe. Add counties, floodedlc, and other appropriate layers to ArcGlobe. Zoom to the study area.

Q22 Describe the bathymetry of the area.

Q23 How would that affect the region's industries?

Hurricane Katrina landed on the western side of Mississippi. Because of this western landing, the storm surge was actually greatest in Hancock and the least in Jackson. Do an individual county analysis. For Hancock, assume the maximum storm surge of 15 feet. For Harrison, assume a storm surge of 11 feet. For Jackson, assume a storm surge of 8 feet.

Reference

Burby, R.J. (2006). Hurricane Katrina and the Paradoxes of Government Disaster Policy: Bringing About Wise Governmental Decisions for Hazardous Areas. *The Annals of the American Academy of Political and Social Sciences,* 604(1), pp. 171–191.

MAKING
SPATIAL
DECISIONS
USING GIS

1

*Hurricane
damage
decisions*

MAKING
SPATIAL
DECISIONS
USING GIS

2

*Hurricane
damage
decisions*

PROJECT 2
Hurricane Wilma storm surge

Scenario

After stalling for several days over Cancun, Mexico, Hurricane Wilma approached the Florida Keys and strengthened to a category 3 storm before making landfall on October 19, 2005, at Key West, Florida. The greatest devastation caused by Wilma was not from the wind but from the storm surge, which was approximately 8 feet. Sixty percent of the homes in Key West were flooded and tens of thousands of cars were submerged.

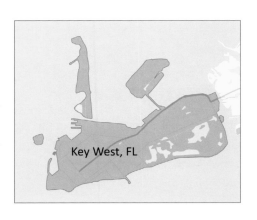

Problem

Most of Wilma's damage was caused by the surge on the morning after the storm. To settle thousands of claims, insurance companies needed maps showing the height of the surge. You are assigned to create maps to assess the total acreage of the different types of land cover that were under water as a result of the Wilma storm surge. You also must map damage to the infrastructure and health-care centers so that restoration efforts can focus on the areas with the greatest need.

Reminder: It helps to divide this large problem into a set of smaller tasks such as the following:

1. Identify the geographic study area.
2. Determine the sequence of steps in your study.
3. Identify the decisions to be made.
4. Develop the information required to make decisions.
5. Identify stakeholders for this issue.

MAKING
SPATIAL
DECISIONS
USING GIS

2

Hurricane
damage
decisions

The questions in this project are both quantitative and qualitative. They identify key points that should be addressed in your analysis and presentation.

Deliverables

We recommend the following deliverables for this exercise:

1. A map showing elevation/bathymetry of Key West with roads and places symbolized.
2. A time series map showing Wilma's path with graphs of wind speed and air pressure.
3. A map of flooded land in Key West after the Wilma storm surge. The map should include a bar graph showing percentage of total flooded land by land-cover type.
4. A map showing infrastructure and health facilities at risk from the storm surge.
5. A table showing various land types that were flooded, measured in acres and square miles.
6. False color composite images and derived land cover created using Landsat Thematic Imagery.

Examine the data

The data for this project are stored in the **\Project2_Wilma\Wilma_data** folder.

Reminder: View the item description to investigate the data. The table below helps you organize this information.

Q1 *Investigate the metadata and complete the following table.*

Layer	Data Type	Publication Information: Who Created the Data?	Time Period Data Are Relevant	Spatial Horizontal Coordinate System	Attribute Values	Resolution for Rasters
elev			1999		Elevation is expressed in meters	1 arc second 0.000369 30 meters
kw_places	Vector	Author created data	2007	Geographic	N/A	N/A

Organize and document your work

Reminder: Be sure to refer to the project 1, Katrina exercise, and your process summary.

1. Set up the proper directory structure.
2. Create a process summary.
3. Document the map.
4. Set the data frame properties for each project. Since none of the data used in this project is projected, set the coordinate system to be UTM NAD 1983 Zone 17N in the Data Frame Properties window. This map projection is appropriate for the Key West region and the unit of measurement is meters. Setting the projection will ensure the most accurate calculations in ArcMap.
5. Set the environments:
   a. Set the working directory.
   b. Set the scratch directory.
   c. Set the Output Coordinate System to Same as Display.
   d. Set the Extent to Same as Layer keywest.
   e. Set the Output Cell size to **30**.
   f. Set the Mask to keywest.

MAKING
SPATIAL
DECISIONS
USING GIS

2

*Hurricane
damage
decisions*

Analysis

An important first step in GIS analysis is to develop a basemap of your study area. Complete deliverable 1 and answer the questions below to orient yourself to the study area.

Deliverable 1: A map showing elevation/bathymetry of Key West with roads and places symbolized.

Q2 *What is the highest elevation shown?*

Q3 *What are the places that are extremely vulnerable to flooding?*

Q4 *What different types of coastal land are represented?*

Deliverable 2: A time series map showing Wilma's path with graphs of wind speed and air pressure.

Reminder: Add states, track_wilma, and keywest and zoom to full extent.

Q5 *Write an analysis listing all the variables and their relationships. For example, the relationship of wind speed to location on land versus water, wind speed to depth of water, etc.*

The next step is to determine what land was flooded by the storm surge and to plot the flooded land by land-cover type. Complete deliverables 3 and 4 and answer the questions to conduct this analysis.

Reminder: The elevation is in meters and the storm surge is given in feet.

Deliverable 3: A map of flooded land in Key West after the Wilma storm surge. The map should include a bar graph showing percentage of total flooded land by land-cover type.

MAKING
SPATIAL
DECISIONS
USING GIS

2

*Hurricane
damage
decisions*

Reclassify the land-cover values as follows:

Old Values	Label	New Values
11	Water	1
21,22,23	Developed	2
31	Barren	3
51,61,71	Scrub/Grass	4
91,92	Wetlands	5

Q6   What type of land was most flooded?

Q7   Describe the flooding of Key West.

Calculate the percentage of land cover by type. [Reminder: PERCENT = (COUNT/Sum of COUNT*100).] You do not want to include the pixels for the land that was not flooded.

Q8   Record the sum of the COUNT field (excluding not flooded land).

Make a bar graph of the percentage of land cover by type.

Q9   What does the graph tell you about the greatest impact of the storm surge?

Q10   Describe the distribution of the flooded areas and how the flood might affect the long-term sustainability of the region.

You also need a map showing what hospitals and other important parts of the Key West infrastructure are under threat from the storm.

Deliverable 4: A map showing infrastructure and health facilities at risk from the storm surge.

Q11 *What is the distribution of infrastructure and health facilities affected by the storm surge and how might you prioritize which damaged elements should be restored first?*

The final piece of the puzzle is to determine the area of the different land types flooded.

Deliverable 5: A table showing various land types that were flooded, measured in acres and square miles.

MAKING
SPATIAL
DECISIONS
USING GIS

2

Hurricane
damage
decisions

Reminder: The field type should be float.
   ACRES = COUNT*900/4046.68
   SQUARE MILES = ACRES/640

Q12 *Complete this table on your worksheet.*

Type	Acres	Sq. Miles
Not Flooded		
Water		
Developed		
Barren		
Scrub/Grass		
Wetlands		

Deliverable 6: False color composite images and derived land cover created using Landsat Thematic Imagery.

Q13 *Write an analysis comparing how different land classes appear in the true and false color composite. Also include in your analysis your thoughts on the accuracy of the land-cover classification as you compared it to both true color, false color, and the aerial photography brought in from ArcGIS Online.*

Presentation

Keep in mind the interests and expertise of your audience as you prepare your presentation. Develop a solution to the original problem and present your results in a compelling way.

Refer to the list of presentation format options in project 1.

MAKING
SPATIAL
DECISIONS
USING GIS

3

Hurricane
damage
decisions

PROJECT 3
On your own

You have worked through a guided activity on the impact of hurricanes on a coastal area and repeated that analysis for another community. In this section you will reinforce the skills you have developed by researching and analyzing a similar scenario entirely on your own. First you must identify your study area and acquire data for your analysis. Several possible storms (and their impact areas) are suggested below. However, if there is another storm that has significant interest to you (perhaps one that has affected your area), download and work with that data.

- New Orleans/Katrina
- Mississippi (or Appalachian Valley)/Camille
- Florida/Camille
- Charleston, South Carolina/Andrew
- Gulf Coast/Rita
- Wilmington, North Carolina/Fran

Refer to your process summary and the preceding module projects if you need help. Here are some basic steps to help you organize your work:

Research

Research the particular event and answer the following questions:

1. What is the area of study?
2. What is the extent of the storm surge or flooding?
3. What were the critical issues of the event?

Obtain the data

Do you have access to baseline data? The Esri Data & Maps Media Kit provides many of the layers of data needed for project work. Be sure to pay particular attention to the source of each data layer and get the latest version. Also use the StreetMap USA database for accurate street data, if you need it. If you do not have access to the Media Kit, you can obtain data from the following sources:

- Census 2000 TIGER/Line Data http://www.esri.com/tiger
- The National Atlas http://www.nationalatlas.gov
- Geospatial One Stop http://gos2.geodata.gov/wps/portal/gos

MAKING
SPATIAL
DECISIONS
USING GIS

3

Hurricane
damage
decisions

Workflow

After researching the problem and obtaining the data, you should do the following:

1. Write a brief scenario.
2. State the problem.
3. Define the deliverables.
4. Examine the data using ArcCatalog.
5. Set the directory structure, start your process summary, and document the map.
6. Decide what you need for the data frame coordinate system and the environments.
   a. What is the best projection for your work?
   b. Do you need to set a cell size or mask?
7. Start your analysis.
8. Prepare your presentation and deliverables.

Always remember to document your work in a process summary.

Presentation

Refer to the list of presentation format options in project 1.

URBAN PLANNING DECISIONS

Introduction

GIS tools can be used to support the planning and development or redevelopment of urban areas. Regional or urban landscape development has historically been a hotbed for the use of GIS. Ian McHarg, the renowned landscape architect, was a pioneer in using map overlays and this overlay planning has seen continual development using GIS (McHarg 1967). Weighted (or sum) overlays are one way to combine layers for GIS planning. However, in weighted overlays, each cell is either in a class or not; there is no way to include a probability of membership. A different way to perform overlays is to use fuzzy logic, which is based not on clear-cut boundaries but rather the imprecision of class boundaries.

Scenarios in this module

- Where to grow in San Diego
- Analyzing urban sprawl in Seattle
- On your own

GIS software required

- ArcGIS Desktop 10 (ArcEditor)
- ArcGIS Spatial Analyst

Student worksheets

The student worksheet files can be found on the Data and Resources DVD.

Project 1: San Diego student sheet
- File name: San_Diego_student_worksheet.doc
- Location: Project1_San_Diego\Documents

Project 2: Seattle student sheet
- File name: Seattle_student_worksheet.doc
- Location: Project2_Seattle\Documents

MAKING
SPATIAL
DECISIONS
USING GIS

1

*Urban
planning
decisions*

PROJECT 1
Where to grow in San Diego

Urban planners have embraced the use of GIS to assist them and involve the public in decision making (Craig, Harris, and Weiner 2002). GIS has also played a role in exploring the connection between public health and urban planning (Saelens, Sallis, and Frank 2003). San Diego urban planners have identified parameters of a potential development scenario for the expansion of the urban area around the Southern California city. They have held community meetings and met with developers to establish four main goals of development that take into consideration proximity to existing development and infrastructure, protection of established parks, and a desire to incorporate aesthetically pleasing views into the landscape.

Scenario

The planners have hired your GIS firm to do a landscape assessment of the land in San Diego County and to provide the planners with a map incorporating the identified parameters. Your firm has determined that a number of maps need to be created in order to do the landscape analysis.

Problem

You first must create distance maps and then convert these maps into fuzzy overlays that can be combined. This analysis would give the planners a clear indication of suitable locations surrounding San Diego for expanding the urban area. In applying GIS to a problem, you must first have a clear understanding of the situation. We find it helpful to answer these four questions that test your understanding and divide the problem into smaller problems that are easier to solve. Record your answers on the worksheet provided.

Q1   **What geographic area are you studying?**

Q2   **What decisions do you need to make?**

Q3   **What information would help you make the decisions?**

Q4   **Who are the key stakeholders for this issue?**
(This step is important. You need to know the audience for your analysis to help decide how to present your results.)

Deliverables

After identifying the problem, you need to envision the kinds of data displays (maps, graphs, and tables) that will provide the solution. We recommend the following deliverables for this exercise:

1. A map of San Diego and vicinity showing the urban area, parks, and major highways.
2. A map of the Euclidean distance around the urban area, parks, and highways plus a classified elevation map with the middle elevations highlighted.
3. A map showing the decision criteria scaled to values ranging from zero to one indicating their strength in the fuzzy membership set.
4. A map showing a fuzzy overlay of the decision criteria plus a model with graphic representation of the fuzzy membership tools used.

MAKING
SPATIAL
DECISIONS
USING GIS

1

Urban
planning
decisions

Tips and tools

Topical instructions are given in the following exercises. If more detailed instructions are needed, ArcGIS Desktop 10 provides these options:

1. Use the Help file to ask a question or look up a keyword, such as a tab, menu option, or function. If you are online, it is better to use the Web-based help option by accessing the ArcGIS Resource Center for an up-to-date version of the help system included with the software.

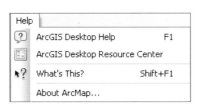

2. The tools can be accessed by using the traditional ArcToolbox or you can use the Search For Tools option found under the Geoprocessing menu.

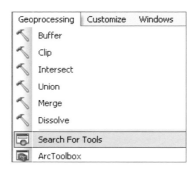

When you search for the tools, an explanation and a link to the tool appear.

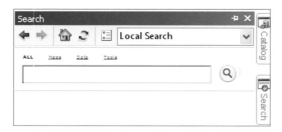

The questions in this project are both quantitative and qualitative. They identify key points that should be addressed in your analysis and presentation.

Examine the data

The next step in your workflow is to identify, collect, and examine the data for the San Diego urban analysis. Here, we have identified and collected the data layers you will need. Explore the data to better understand both the raster and vector feature classes in this exercise. In order to explore the data, you need to access the metadata associated with each feature class. Thoroughly investigate the data layers to understand how they will help you address the problem. The spatial coordinate system, the resolution of the data, and the attribute data are all important pieces of information about a feature class.

MAKING
SPATIAL
DECISIONS
USING GIS

1

Urban
planning
decisions

1. Open ArcMap. (For these exercises, the Getting Started dialog box is not needed. Select the "Do not show this dialog in the future" option.)

2. Add Data by connecting to the \Project1_San_Diego\San_Diego_data folder. The data folder contains the San_Diego geodatabase, which holds four feature classes and a raster. Add the raster elevation and the features San Diego, urban, parks, and highways.

There are different metadata styles that control how you view an item's description. The metadata data style "ISO 19139 Metadata Implementation Specification" supports formal metadata and allows the complete metadata to appear in the Item Description window. This setting must be made in the stand-alone ArcCatalog software.

3. Open ArcCatalog and in the Customize menu choose ArcCatalog Options. Select the Metadata tab and select "ISO 19139 Metadata Implementation Specification" from the Item Description menu.

4. Close ArcCatalog.

Q5 *View the item descriptions for these features and complete the following table on your worksheet.*

Layer	Publication Information: Who Created the Data?	Time Period Data Are Relevant	Spatial Horizontal Coordinate System	Data Type	Resolution for Rasters	Attribute Values
elevation	USGS	2009			30 meters	N/A
San_Diego			NAD_1983_UTM_Zone_11N		N/A	

5. Close ArcMap.

Now that you have explored the available data, you are almost ready to begin your analysis. First you need to start a process summary, document your project, and set the project environments.

Organize and document your work

The following preliminary steps are essential to a successful GIS analysis.

MAKING
SPATIAL
DECISIONS
USING GIS

1

*Urban
planning
decisions*

Examine the directory structure

The next phase in a GIS project is to carefully keep track of the data and your calculations. You will work with a number of different files and it is important to keep them organized so you can easily find them. The best way to do this is to have a folder for your project that contains a data folder. For this project, the folder named **\Project1_San_Diego\San_Diego_data** will be your project folder. Make sure that it is stored in a place where you have write access. You can store your data inside the results folder. The results folder already contains an empty geodatabase named **\San_Diego_results**. Save your map documents inside the **\San_Diego_results** folder.

```
☐ 🗁 EsriPress
    ☐ 🗁 MakingSpatialDecisions
        ☐ 🗁 Urban_planning_decisions
            ☐ 🗁 Project1_San_Diego
                ⊞ 🗁 Documents
                ⊞ 🗁 San_Diego_data
                ⊞ 🗁 San_Diego_results
            ☐ 🗁 Project2_Seattle
                ⊞ 🗁 Documents
                ⊞ 🗁 Seatlle_results
                ⊞ 🗁 Seattle_data
```

Create a process summary

The process summary is simply a list of the steps you used to do your analysis. We suggest using a simple text document for your process summary. Keep adding to it as you do your work to avoid forgetting any steps. The list below shows an example of the first few entries in a process summary:

1. Explore the data.
2. Produce a basemap of San Diego's urban area, parks, and highways. Label and symbolize.
3. Calculate a Euclidean distance around Urban, Parks, and Highways.

Document the map

1. **Open ArcMap and save the map document as** SanDiego1. **Save it in the** \San_Diego_results **folder.**

You need to add descriptive properties to every map document you produce. Use the same descriptive properties for every map document in the module or individualize the documentation from map to map. You can access the Map Document Properties from the File menu. The Document Properties dialog box allows you to add a title, summary, description, author, credits, tags, and hyperlink base. After writing descriptive properties, be sure to select the Pathnames check box, which makes ArcMap store relative paths to all of your data sources. Storing relative paths allows ArcMap to automatically find all relevant data if you move your project folder to a new location or computer.

Set the environments

MAKING
SPATIAL
DECISIONS
USING GIS

1

Urban
planning
decisions

In GIS analysis, you will often get data from several sources and these data may be in different coordinate systems and/or map projections. When using GIS to perform area calculations, you would like your result to be in familiar units, such as miles or kilometers. Data in an unprojected geographic coordinate system have units of decimal degrees, which are difficult to interpret. Thus, your calculations will be more meaningful if all the feature classes involved are in the same map projection. Fortunately, ArcMap can do much of this work for you if you set certain environment variables and data frame properties. In this section, you will learn how to change these settings. To display your data correctly, you will need to set the coordinate system for the data frame. When you add data with a defined coordinate system, ArcMap will automatically set the data frame's projection to match the data. If you add subsequent layers that have a coordinate system different from the data frame, they are automatically projected on-the-fly to the data frame's coordinate system.

1. From the View menu, choose Data Frame Properties. Click the Coordinate System tab. Select Predefined, Projected Coordinate Systems, UTM, NAD 1983, Zone 11N, and click OK. This sets the projection for the active data frame.

2. From the Geoprocessing menu, choose Environment. Remember that the environment settings apply to all the functions within the model. The analysis environment includes the workspace where the results will be placed, and the extent, cell size, and coordinate system for the results.

3. Expand workspace. By default, inputs and outputs are placed in your current workspace, but you can redirect the output to another workspace such as your results folder. Set the Current Workspace as \Project1_San_Diego\San_Diego_data\San_Diego.gdb. Set the Scratch Workspace as \Project1_San_Diego\San_Diego_results\San_Diego_results.gdb.

4. For Output Coordinate System, select Same as Display.

5. Set the Processing Extent to be Same as Layer San_Diego.

6. Expand the Raster Analysis Settings and set the Cell size to be the same as elevation.

You also want to limit your analysis to the county of San Diego. This is accomplished by using an analysis mask. The mask identifies those locations within the analysis extent that will be included when using a tool.

7. Set the Mask to San_Diego from the San_Diego geodatabase.

8. Click OK and save the project again as SanDiego1.

Analysis

Before beginning the analysis, it is a good idea to review the urban planners' decision parameters. They had four main goals for the new urban development:

- Proximity to existing urban areas.
- Proximity to highways.
- Protection of parks.
- Aesthetically pleasing views of landscape. Middle-range elevations provide the most aesthetically pleasing views.

Once you have examined the data, completed map documentation, and set the environments, you are ready to begin the analysis and build the displays to address the problem. A good place to start any GIS analysis is to produce a basemap to better understand the distribution of features in the geographic area you are studying. First, you will prepare a basemap of San Diego showing urban area, parks, and highways.

STEP 1: Create a basemap of San Diego

1. Create a basemap of San Diego by adding San_Diego, parks, urban, and highways from the San_Diego geodatabase.

2. Symbolize the layers appropriately and import the highways layer file.

3. Save the map document as SanDiego1.

4. Save the map document again as SanDiego2. (When you save the map document again as SanDiego2, it correctly saves documentation, the Data Frame projection, and the Environment settings. This saves you from redoing these variables for other deliverables.)

Deliverable 1: A map of San Diego and vicinity showing the urban area, parks, and major highways.

STEP 2: Calculate Euclidean distance around Urban, Parks, and Highways

Euclidean distance is defined as the distance that you would measure with a ruler in a straight line. The Euclidean Distance tool used in this analysis calculates straight-line distance from the polygon features and converts the results internally into a raster.

1. Open San Diego2 and rename the data frame Urban Distance. Copy this data frame and paste it twice, renaming the new data frames Parks Distance and Highway Distance, respectively.

Reminder: To make a data frame active, right-click its name in the table of contents and select Activate.

A. Euclidean distance around Urban

1. In the Urban Distance data frame, remove parks and highways.

2. Calculate the Euclidean distance by using urban as the feature source data.

3. Name the output distance raster EucDist_urban and save in \San_Diego_results.gdb.

4. Classify manually using appropriate break values such as 15,000; 30,000; 45,000; 60,000; and the highest value given.

B. Euclidean distance around Parks

1. Activate the Parks Distance data frame and remove urban and highways.

2. Calculate the Euclidean distance by using parks as the feature source data.

3. Name the output distance raster EucDist_park and save in \San_Diego_results\San_Diego_results.gdb.

4. Classify manually using appropriate break values such as 5,000; 10,000; 15,000; 30,000; and the highest value given.

C. Euclidean distance around Highways

1. Activate the Highways Distance data frame, remove urban and parks.

2. Calculate the Euclidean distance by using highways as the feature source data.

MAKING
SPATIAL
DECISIONS
USING GIS

1

*Urban
planning
decisions*

3. Name the output distance raster EucDist_high **and save in** \San_Diego_results\San_Diego_results.gdb.

4. Classify manually using appropriate break values such as 4,000; 8,000; 12,000; 16,000; and the highest value given.

STEP 3: Display elevation with middle elevations highlighted

MAKING
SPATIAL
DECISIONS
USING GIS

1

Urban
planning
decisions

1. Copy the data frame Highway Distance and paste it. Rename it as Optimal Elevation. **In the Optimal Elevation data frame, remove highways,** EucDist_high **and add elevation.**

Q6 ***Write a paragraph describing the spatial display of elevation in San Diego. Include the following details in your paragraph:***
   - The lowest and highest elevations and which shades of gray represent them.
   - Which shades of gray represent the middle elevations? (Hint: You can see this more clearly if you classify the elevation by quantities.)

2. Classify elevation into ten equal interval classes. Highlight the middle elevations of approximately 1,000 to 1,200.

3. Create a presentation layout showing all four data frames. Use proper cartographic principles.

4. Save the map document as SanDiego2.

5. Save the map document again as SanDiego3.

Deliverable 2: A map of the Euclidean distance around the urban area, parks, and highways plus a classified elevation map with the middle elevations highlighted.

STEP 4: Use the Fuzzy Membership tool

In standard weighted overlays, each pixel in a raster contributes to a calculation completely or not all; that is, it has a value of either 1 or 0. This means that you have criteria that let you say with certainty that a particular location counts or it does not. However, since real situations are not black or white, there is often some probability that a particular location should count. Fuzzy logic offers us a way to include this variation in our calculations.

In this part of the exercise we will use the Fuzzy Membership tool to scale the input raster data sets so that each pixel has a value ranging from zero to one, indicating the strength of the raster as a part of a set. (Think of this like probability.) A value of one means the raster is a full member of the fuzzy set (and would have 100 percent weight in a calculation), and the value can decrease to zero, meaning that pixel is not a member of the fuzzy set and has a 0 percent weight in a calculation.

The Fuzzy Membership tool has various membership functions. You can decide which function to use based on which best describes the situation, depending on the phenomenon being modeled.

In this exercise FuzzySmall is the transformation function used to scale the input raster datasets of Euc_high and Euc_urban. The Fuzzy Small transformation function is used when the smaller input values are more likely to be a member of the set (distances closer to both urban and highways).

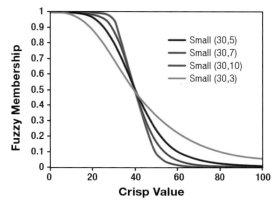

This figure shows the FuzzySmall function. Note that low "crisp values" (raster values) have high membership. This means that pixels with a small crisp value will be much more likely to contribute to the calculation.

Source: Esri Help file.

The Fuzzy Large transformation function is used when the larger input values are more likely to be a member of the set (distance from parks).

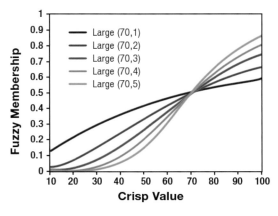

This figure shows the FuzzyLarge function. Note that high "crisp values" (raster values) have high membership. This means that pixels with a large crisp value will be much more likely to contribute to the calculation.

Source: Esri Help file.

The Fuzzy Near transformation function is most useful if membership is near a specific value. The function is defined by a midpoint defining the center of the set, identifying definite membership, and therefore being assigned a 1. As values move from the midpoint, in both the positive

137

and negative directions, membership decreases until it reaches 0, indicating no membership. The spread defines the width and character of the transition zone.

MAKING
SPATIAL
DECISIONS
USING GIS

1

Urban
planning
decisions

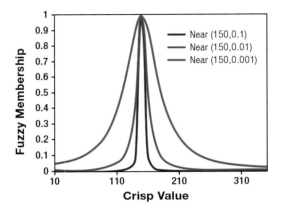

Fuzzy Near and Fuzzy Gaussian can be similar, depending on the specified parameters. The Fuzzy Near function generally decreases at a faster rate, with a more narrow spread, than the Fuzzy Gaussian function and is therefore used when the values very near the midpoint are more likely to be a member of the set.

Source: Esri Help file.

A. Fuzzy Membership Small for Urban Distance (using Small membership favors the goal of "closeness to existing development")

1. Open SanDiego3.

2. Activate the Urban Distance data frame.

3. Use the Fuzzy Membership tool with EucDist_urban as the input raster and FuzzySM_urban as the output raster saved in San_Diego_results.gdb.

4. Select Small as the Membership type.

5. Remove EucDist_urban.

6. Using the Identify tool, verify that geographic areas closer to the urban area have a larger value (closer to 1) than geographic areas away from the urban area (values closer to 0).

B. Fuzzy Membership Small for Highway Distance (using Small membership tool favors the goal of "closeness to existing infrastructure")

1. Activate the Highway Distance data frame.

2. Use the Fuzzy Membership tool with EucDist_high as the input raster and FuzzySm_highway as the output raster saved in San_Diego_results.gdb.

3. Select Small as the Membership type.

4. Remove EucDist_high.

5. Using the Identify tool, verify that geographic areas closer to the highways have a larger value (closer to 1) than geographic areas away from the highways (values closer to 0).

C. Fuzzy Membership Large for Parks (using the Large membership favors the goal of "protection of open areas and parks")

1. Activate the Parks Distance data frame.

2. Use the Fuzzy Membership tool with EucDist_park as the input raster and FuzzyLg_parks as the output raster saved in San_Diego_results.gdb.

3. Select Large as the Membership type.

4. Remove EucDist_park.

5. Using the Identify tool, verify that geographic areas closer to the parks have lower values (closer to 0) than geographic areas away from the parks (values closer to 1).

D. Fuzzy Membership Near for Elevation (using Near membership favors the selection of middle elevation values, which satisfies the goal of "incorporating aesthetically pleasing views")

1. Activate the Optimal Elevation data frame.

2. Use the Fuzzy Membership tool with elevation as the input raster and FuzzyNear_elev as the output raster saved in San_Diego_results.gdb.

3. Select Near as the Membership type and 0.0001 as the Spread value.

4. Using the Identify tool, verify that geographic areas having midrange elevations have higher values (closer to 1) and areas with high or low elevations have lower values (closer to 0).

Q7 What are the middle elevations?

5. Create a presentation layout showing all four data frames. Use proper cartographic principles.

6. Save the map document as SanDiego3.

7. Save the map document again as SanDiego4.

MAKING
SPATIAL
DECISIONS
USING GIS

Urban
planning
decisions

Deliverable 3: A map showing the decision criteria features scaled to values ranging from zero to one, indicating their strength in the fuzzy membership set.

STEP 5: Fuzzy Overlay AND with model

Now that you have created the fuzzy layers for the decision criteria, you need to combine them into a single overlay that will highlight the best areas for additional development.

Combine fuzzy memberships using Fuzzy Overlay AND

1. Open SanDiego4 and rename the Fuzzy Small Urban data frame as Fuzzy Overlay.

2. Copy FuzzySm-highway from the Fuzzy Small Highways data frame and paste it in the Fuzzy Overlay data frame. Delete the Highway Distance data frame.

3. Copy FuzzyLg_parks and paste it in the Fuzzy Overlay data frame. Delete the Parks Distance data frame.

4. Copy FuzzyNear_elev and paste it in the Fuzzy Overlay data frame. Delete the Optimal Elevation data frame.

5. Use the Fuzzy Overlay tool to combine the fuzzy membership rasters. Select AND for the Overlay type. Input the following rasters:
    *   FuzzyNear_elev
    *   FuzzyLg_parks
    *   FuzzySm_highway
    *   FuzzySm_urban

6. Name the output raster overand and save it in San_Diego_results.gdb.

7. Construct a graphic representation of the process you have used to be included on your layout or as a stand-alone representation of your workflow. You may use ModelBuilder or a stand-alone graphics or presentation program to do this step. To use ModelBuilder:
    a. Activate the ModelBuilder window from the standard toolbar.
    b. Add the following files to the model window:
        i.. EucDist_urban
        ii.. EucDist_high
        iii. EucDist_park
        iv. Elevation

   c. Drag the Fuzzy Membership tool to the ModelBuilder window and connect to the Input rasters. Give each process the correct name (Small, Large, Near) and name the output files appropriately.

   d. Drag the Fuzzy Overlay tool to the ModelBuilder window, connect all the output files, and name the overlay.

   e. The model can be exported and added to the layout as a graphic or can be used as a stand-alone diagram of your workflow.

8. Create a presentation layout. Use proper cartographic principles. Use graphics to show the areas best suited for urban development.

9. Save the map document as SanDiego4.

MAKING
SPATIAL
DECISIONS
USING GIS

1

Urban
planning
decisions

Deliverable 4: A map showing the fuzzy overlay of the decision criteria features plus a model with graphic representation of the fuzzy membership tools used.

Once your analysis is complete, you still need to develop a solution to the original problem and present your results in a compelling way to the urban planners in this particular situation. The presentation of your various data displays must explain what they show and how they help solve the problem.

Presentation

There are many ways, ranging from simple to advanced, that you can use to prepare a presentation. Whatever method of presentation you choose has to include a report documenting your analysis and addressing the fuzzy membership and fuzzy overlay types. You must explain the spatial patterns you see and describe the implications of your calculations and analysis for this problem. As you work through this book, try to use different types of presentation media. Pick the presentation medium that best fits your audience. Remember that your audience probably lacks your in-depth knowledge of GIS, so you will need to communicate your results in a way they will be able to understand and use.

Listed below are various presentation formats:
- Create a text document with inserted maps.
- Show your findings in a digital slide presentation.
- Use ArcGIS Explorer Desktop, which is a free, downloadable GIS viewer that provides an easy way to explore, visualize, share, and present geographic information. This software can be downloaded at:

    http://www.esri.com/software/arcgis/explorer/download.html

MAKING
SPATIAL
DECISIONS
USING GIS

1

*Urban
planning
decisions*

- Embed interactive maps in your text documents. These maps can also be shared with others. The links below provide information about this tool:

    http://www.esri.com/software/mapping_for_everyone/index.html

    http://help.arcgis.com/en/webapi/javascript/arcgis/index.html

- Create layer packages or map packages and share your maps with your classmates.
- Use ArcGIS Explorer Online to produce an interactive online geospatial presentation that can be shared: http://www.arcgis.com/home/

References

Craig, W., T. Harris, and D. Weiner (2002). *Community participation and geographic information systems,* Taylor and Francis, New York.

McHarg, Ian (1967). *Design with Nature.* John Wiley and Sons, New York.

Saelens, B., J. Sallis, and L. Frank (2003). Environmental correlates of walking and cycling: Findings from the transportation, urban design, and planning literatures. *Annals of Behavioral Medicine,* 25:2, pp. 80–91.

PROJECT 2
Analyzing urban sprawl in Seattle

MAKING
SPATIAL
DECISIONS
USING GIS

2

*Urban
planning
decisions*

Scenario

A regional council of urban planners has identified development scenario parameters for the area around Seattle. The council wants to slow down urban sprawl by identifying areas close to existing development and infrastructure and away from parks. The urban planners, however, would like to incorporate aesthetically pleasing views into the landscape by focusing on developable land with a midrange elevation. The council has hired your GIS firm to do a landscape assessment of the land around Seattle and to provide the planners with a map incorporating the identified parameters.

143

Problem

A number of maps need to be created for the landscape analysis. You first will create distance maps and then convert these maps into fuzzy overlays that can then be combined. This analysis would then give the urban planners a clear indication of prime places surrounding Seattle to extend the urban area.

MAKING
SPATIAL
DECISIONS
USING GIS

2

*Urban
planning
decisions*

Reminder: It helps to divide this large problem into a set of smaller tasks, such as the following:

- Identify the geographic study area.
- Determine the sequence of steps in your study.
- Identify the decisions to be made.
- Develop the information required to make decisions.
- Identify stakeholders for this issue.

The questions in this project are both quantitative and qualitative. They identify key points that should be addressed in your analysis and presentation.

Deliverables

We recommend the following deliverables for this exercise:

1. A map of Seattle and the surrounding King County showing urban areas, parks, and major highways.
2. A map of the Euclidean distance around the urban area, parks, and highways plus a classified elevation map with the middle elevations highlighted.
3. A map showing the decision criteria scaled to values ranging from zero to one, indicating their strength in a membership set.
4. A map showing the fuzzy overlay of the decision criteria features plus a model with graphic representation of the fuzzy membership tools used.

Examine the data

The data for this project are stored in the **Project2_Seattle\Seattle_data** folder.

Reminder: View the item description to investigate the data. The table helps you organize this information.

Q1 *Investigate the metadata and complete the following table on your worksheet.*

Layer	Publication Information: Who Created the Data?	Time Period Data Are Relevant	Spatial Horizontal Coordinate System	Data Type	Resolution for Rasters	Attribute Values
elev	USGS	2009				N/A
King	Esri		UTM NAD 1983 Zone 10N		N/A	

Organize and document your work

Be sure to refer to the project 1 San Diego exercise and your process summary.

1. Set up the proper directory structure.
2. Create a process summary.
3. Document the map.
4. Set the environments:
    a. Set the Data Frame Coordinate System to be UTM NAD 1983 Zone 10.
    b. Set the working directory.
    c. Set the scratch directory.
    d. Set the Output Coordinate System to Same as Display.
    e. Set the Extent to Same as King.
    f. Set the Output Cell Size to be the same as elevation.
    g. Set the Mask to be King.

Analysis

An important first step in GIS analysis is to develop a basemap of your study area. Complete deliverable 1 and answer the questions below to orient yourself to the study area.

Deliverable 1: A map of Seattle and the surrounding King County showing the urban area, parks, and major highways.

Reminder: Calculate Euclidean distance around Urban, Parks, and Highways. Also display a classified elevation with middle elevations highlighted.

MAKING
SPATIAL
DECISIONS
USING GIS

2

Urban
planning
decisions

Q2 *Write a paragraph describing the spatial display of elevation in Seattle.*

Include the following details in your paragraph:

- The lowest and highest elevations and which shades of gray represent them.
- What shades of gray represent the middle elevation? (Hint: You can see this more clearly if you classify the elevation by equal interval.)

Deliverable 2: A map of the Euclidean distance around the urban area, parks, and highways plus a classified elevation map with the middle elevations highlighted.

Q3 *What is the middle elevation?*

Reminder: Use the Fuzzy Membership tool to change Euclidean distances and elevation into scaled values.

Remember that when using the Near Fuzzy Membership tool set the spread to be 0.0001.

Deliverable 3: A map showing the decision criteria features scaled to values ranging from zero to one, indicating their strength in the fuzzy membership set.

Reminder: Use the scaled rasters to calculate a fuzzy overlay.

Deliverable 4: A map showing the fuzzy overlay of the decision criteria features plus a model with graphic representation of the fuzzy membership tools used.

Presentation

Keep in mind the interests and expertise of your audience as you prepare your presentation. Develop a solution to the original problem and present your results in a compelling way.

Refer to the list of presentation format options in project 1.

MAKING
SPATIAL
DECISIONS
USING GIS

3

Urban
planning
decisions

PROJECT 3
On your own

You have worked through a guided activity on urban planning in San Diego and you have repeated the analysis with Seattle data. In this section, you will reinforce your skills by researching and analyzing a similar scenario entirely on your own. First, you must identify your study area and acquire the data for your analysis. You may want to do a local urban area.

Refer to your process summary and the preceding module projects if you need help. Here are some basic steps to help you organize your work.

Research

Research the problem and answer the following questions:

1. What is the area of study?
2. What data are available?

Obtain the data

Do you have access to baseline data? The Esri Data & Maps Media Kit provides many of the layers of data that are needed for project work. Be sure to pay particular attention to the source of data and get the latest version. If you do not have access to the Media Kit, you can obtain data from the following sources:

*   Census 2000 TIGER/line data http://www.esri.com/tiger
*   Geospatial One Stop http://gos2.geodata.gov/wps/portal/gos
*   The National Atlas http://www.nationalatlas.gov
*   U.S. Geological Survey Seamless Data Warehouse http://seamless.usgs.gov

Workflow

After researching the problem and obtaining the data, you should do the following:

1. Write a brief scenario.
2. State the problem.
3. Define the deliverables.
4. Examine the metadata.
5. Set the directory structure, start your process summary, and document the map.
6. Decide what you need for the data frame coordinate system and the environments.
   a. What is the best projection for your work?
   b. Do you need to set a cell size or mask?
7. Start your analysis.
8. Prepare your presentation and deliverables.
9. Always remember to document your work in a process summary.

Presentation

Refer to the list of presentation format options in project 1.

MAKING
SPATIAL
DECISIONS
USING GIS

3

Urban
planning
decisions

Data sources

MakingSpatialDecisions\Hazardous_emergency_decisions\Project1_Springfield\Springfield_data

Data sources include:

\data\Springfield.gdb\aerial, data available from the U.S. Geological Survey

\data\Springfield.gdb\additional_layers\bldg, courtesy of County of Fairfax, VA

\data\Springfield.gdb\additional_layers\counties, from Esri Data & Maps 2006,
    courtesy of ArcWorld Supplement

\data\Springfield.gdb\additional_layers\gschools, from Esri Data & Maps 2006,
    courtesy of U.S. Geological Survey-GNIS

\data\Springfield.gdb\additional_layers\route1, created by the author

\data\Springfield.gdb\additional_layers\route2, created by the author

\data\Springfield.gdb\additional_layers\route3, created by the author

\data\Springfield.gdb\network\usastreets, from Esri Data & Maps 2006, courtesy of Esri

\images\springfield_1.tif, created by the author

MakingSpatialDecisions\Hazardous_emergency_decisions\Project2_Mecklenburg
\Mecklenburg_data

Data sources include:

\data\Mecklenburg.gdb\aerial, data available from the U.S. Geological Survey

\data\Mecklenburg.gdb\AdditionalLayers\blkgrp, from Esri Data & Maps 2006, courtesy of Tele Atlas,
    U.S. Census, Esri (Pop2005 field)

\data\Mecklenburg.gdb\AdditionalLayers\county, from Esri Data & Maps 2006, courtesy of ArcUSA,
    U.S. Census, Esri (Pop2005 field)

\data\Mecklenburg.gdb\AdditionalLayers\detour1, created by the author

\data\Mecklenburg.gdb\AdditionalLayers\detour2, created by the author

\data\Mecklenburg.gdb\AdditionalLayers\detour3, created by the author

\data\Mecklenburg.gdb\AdditionalLayers\detour4, created by the author

\data\Mecklenburg.gdb\AdditionalLayers\detour5, created by the author

\data\Mecklenburg.gdb\AdditionalLayers\gschools, from Esri Data & Maps 2006,
    courtesy of U.S. Geological Survey-GNIS

\data\Mecklenburg.gdb\Network\usastreets, from Esri Data & Maps 2006, courtesy of Esri

\images\mecklenburg.tif, created by the author

MakingSpatialDecisions\Demographic_decisions\Project1_Chicago\Chicago_data

Data sources include:

\data\dt_dec_2000_sf4_u_data001_1.xls, courtesy of U.S. Census

\data\Chicago.gdb\chicago, from Esri Data & Maps, 2006, courtesy of National Atlas of the United States

\data\Chicago.gdb\county, from Esri Data & Maps 2006, courtesy of Esri, derived from Tele Atlas,
    U.S. Census, Esri (Pop2005 field)

\data\Chicago.gdb\tracts_00, from Esri Data & Maps 2006, courtesy of Tele Atlas, U.S. Census,
    Esri (Pop2005 field)

\data\Chicago.gdb\tracts_90, from Esri Data & Maps, 1999, courtesy of U.S. Census

\images\map.tif, created by the author

MakingSpatialDecisions\Demographic_decisions\Project2_DC\DC_data
Data sources include:

\data\dt_dec_2000_sf3_u_data1.xls, courtesy of U.S. Census

\data\DC.gdb\dc, from Esri Data & Maps 2006, courtesy of Esri, derived from Tele Atlas, U.S. Census, Esri
    (Pop2005 field)

\data\DC.gdb\dtl_water, from Esri Data & Maps 2006, courtesy of U.S. Geological Survey, Esri

\data\DC.gdb\mjr_hwys, from Esri Data & Maps, 2006, courtesy of Esri

\data\DC.gdb\tracts_00, from Esri Data & Maps, 2006, courtesy of Tele Atlas, U.S. Census,
    Esri (Pop2005 field)

\data\DC.gdb\tracts_90, from Esri Data & Maps, 1999, courtesy of U.S. Census

\images\map.tif, created by the author

MakingSpatialDecisions\Law_enforcement_decisions\Project1_Houston\Houston_data
Data sources include:

\data\aug06.xls, courtesy of Houston Police Department

\data\police_stations.txt, courtesy of Houston Police Department

\data\Houston.gdb\blkgrp, from Esri Data & Maps, 2006, courtesy of Tele Atlas, U.S. Census,
    Esri (Pop2005 field)

\data\Houston.gdb\highways, from Esri Data & Maps, 2006, courtesy of Tele Atlas

\data\Houston.gdb\houston, from Esri Data & Maps, 2006, courtesy of U.S. Census

\data\Houston.gdb\places, from Esri Data & Maps, U.S. Census

\data\Houston.gdb\schools, from Esri Data & Maps, USGS-GNIS

\data\Houston.gdb\usa_sts, from Esri Data & Maps, 2006, courtesy of Esri

\images\intro.tif, created by the author

MakingSpatialDecisions\Law_enforcement_decisions\Project2_Lincoln\Lincoln_data
Data sources include:

\data\Lincoln.gdb\police_stations.txt, courtesy of Lincoln Police Department

\data\Lincoln.gdb\UCR CODES.txt, courtesy of Lincoln Police Department

\data\Lincoln.gdb\aug_06, courtesy of Lincoln Police Department

\data\Lincoln.gdb\blkgrp, from Esri Data & Maps, 2006, courtesy of Tele Atlas, U.S. Census,
    Esri (Pop2005 field)

\data\Lincoln.gdb\lincoln, from Esri Data & Maps, 2006, courtesy of U.S. Census

\data\Lincoln.gdb\places, from Esri Data & Maps, U.S. Census

\data\Lincoln.gdb\usa_streets, from Esri Data & Maps, 2006, courtesy of Esri

\images\intro.tif, created by the author

MakingSpatialDecisions\Hurricane_damage_decisions\Project1_Katrina\Katrina_data

Data sources include:

\data\Landsat5\Sept_2005, data available from the U.S. Geological Survey

\data\Katrina.gdb\airports, from Esri Data & Maps, 2006, courtesy of National Atlas of the United States

\data\Katrina.gdb\churches, from Esri Data & Maps, 2006, courtesy of U.S. Geological Survey – GNIS

\data\Katrina.gdb\counties, data available from the U.S. Geological Survey

\data\Katrina.gdb\elev, data available from the U.S. Geological Survey

\data\Katrina.gdb\hospitals, from Esri Data & Maps, 2006, courtesy of U.S. Geological Survey – GNIS

\data\Katrina.gdb\islands, data available from the U.S. Geological Survey

\data\Katrina.gdb\landcover, data available from the U.S. Geological Survey

\data\Katrina.gdb\places, from Esri Data & Maps, U.S. Census

\data\Katrina.gdb\railroads, from Esri Data & Maps, 2006,
    courtesy of U.S. Bureau Transportation Statistics

\data\Katrina.gdb\rivers, from Esri Data & Maps, 2006, courtesy of U.S. Geological Survey, Esri

\data\Katrina.gdb\states, from Esri Data & Maps, 2006, courtesy of Esri, derived from Tele Atlas,
    U.S. Census, Esri (Pop2005 field)

\data\Katrina.gdb\track_katrina, courtesy of National Oceanic Atmospheric Administration

\data\Katrina.gdb\usa_streets, from Esri Data & Maps, 2006, courtesy of Esri

\data\Katrina.gdb\water, from Esri Data & Maps, 2006, courtesy of U.S. Geological Survey, Esri

\images\intro.tif, created by the author

MakingSpatialDecisions\Hurricane_damage_decisions\Project2_Wilma\Wilma_data

Data sources include:

\data\Landsat5\Jan_06, data available from the U.S. Geological Survey

\data\Wilma.gdb\airports, from Esri Data & Maps, 2006, courtesy of National Atlas of the United States

\data\Wilma.gdb\elev, data available from the U.S. Geological Survey

\data\Wilma.gdb\gchurch, from Esri Data & Maps, 2006, courtesy of U.S. Geological Survey – GNIS

\data\Wilma.gdb\ghospitals, from Esri Data & Maps, 2006, courtesy of U.S. Geological Survey – GNIS

\data\Wilma.gdb\gschools, from Esri Data & Maps, 2006, courtesy of U.S. Geological Survey – GNIS

\data\Wilma.gdb\keywest, from Esri Data & Maps, 2006, courtesy of Esri, derived from Tele Atlas,
    U.S. Census, Esri (Pop2005 field)

\data\Wilma.gdb\places, created by the author

\data\Wilma.gdb\landcover, data available from the U.S. Geological Survey

\data\Wilma.gdb\track_wilma, courtesy of National Oceanic Atmospheric Administration

\data\Wilma.gdb\states, from Esri Data & Maps, 2008, courtesy of Esri, derived from Tele Atlas,
    U.S. Census, Esri (Pop2007 field)

\data\Wilma.gdb\usa_streets, from Esri Data & Maps, 2006, courtesy of Esri

\images\intro.tif, created by the author

MakingSpatialDecisions\Urban_planning_decisions\Project1_San_Diego\San_Diego_data

Data sources include:

\data\San_Diego.gdb\elevation, data available from the U.S. Geological Survey

\data\San_Diego.gdb\highways, from Esri Data & Maps, 2006, courtesy of Esri

\data\San_Diego.gdb\parks, from Esri Data & Maps, 2006, courtesy of National Park Service, ArcUSA,
    Tele Atlas

\data\San_Diego.gdb\San_Diego, from Esri Data & Maps, 2008, courtesy of Esri, derived from Tele Atlas,
    U.S. Census, Esri (Pop2007 field)

\data\San_Diego.gdb\urban, from Esri Data & Maps, 2006, courtesy of U.S. Census

\images\intro.tif, created by the author

Data sources

MakingSpatialDecisions\Urban_planning_decisions\Project2_Seattle\Seattle_data

Data sources include:

\data\Seattle.gdb\elev, data available from the U.S. Geological Survey

\data\Seattle.gdb\Highways, from Esri Data & Maps, 2006, courtesy of Esri

\data\Seattle.gdb\King, from Esri Data & Maps, 2008, courtesy of Esri, derived from Tele Atlas,
    U.S. Census, Esri (Pop2007 field)

\data\Seattle.gdb\Parks, from Esri Data & Maps, 2006, courtesy of National Park Service, ArcUSA,
    Tele Atlas

\data\Seattle.gdb\Urban, from Esri Data & Maps, 2006, courtesy of U.S. Census

\images\intro.tif, created by the author

Data license agreement

Important: Read carefully before opening the sealed media package.

Environmental Systems Research Institute, Inc. (Esri) is willing to license the enclosed data and related materials to you only upon the condition that you accept all of the terms and conditions contained in this license agreement. Please read the terms and conditions carefully before opening the sealed media package. By opening the sealed media package, you are indicating your acceptance of the Esri License Agreement. If you do not agree to the terms and conditions as stated, then Esri is unwilling to license the data and related materials to you. In such event, you should return the media package with the seal unbroken and all other components to Esri.

Esri License Agreement

This is a license agreement, and not an agreement for sale, between you (Licensee) and Environmental Systems Research Institute, Inc. (Esri). This Esri License Agreement (Agreement) gives Licensee certain limited rights to use the data and related materials (Data and Related Materials). All rights not specifically granted in this Agreement are reserved to Esri and its Licensors.

Reservation of Ownership and Grant of License: Esri and its Licensors retain exclusive rights, title, and ownership to the copy of the Data and Related Materials licensed under this Agreement and, hereby, grant to Licensee a personal, nonexclusive, nontransferable, royalty-free, worldwide license to use the Data and Related Materials based on the terms and conditions of this Agreement. Licensee agrees to use reasonable effort to protect the Data and Related Materials from unauthorized use, reproduction, distribution, or publication.

Proprietary Rights and Copyright: Licensee acknowledges that the Data and Related Materials are proprietary and confidential property of Esri and its Licensors and are protected by United States copyright laws and applicable international copyright treaties and/or conventions.

Permitted Uses: Licensee may install the Data and Related Materials onto permanent storage device(s) for Licensee's own internal use.

Licensee may make only one (1) copy of the original Data and Related Materials for archival purposes during the term of this Agreement unless the right to make additional copies is granted to Licensee in writing by Esri.

Licensee may internally use the Data and Related Materials provided by Esri for the stated purpose of GIS training and education.

Uses Not Permitted: Licensee shall not sell, rent, lease, sublicense, lend, assign, time-share, or transfer, in whole or in part, or provide unlicensed Third Parties access to the Data and Related Materials or portions of the Data and Related Materials, any updates, or Licensee's rights under this Agreement. Licensee shall not remove or obscure any copyright or trademark notices of Esri or its Licensors.

Term and Termination: The license granted to Licensee by this Agreement shall commence upon the acceptance of this Agreement and shall continue until such time that Licensee elects in writing to discontinue use of the Data or Related Materials and terminates this Agreement. The Agreement shall automatically terminate without notice if Licensee fails to comply with any provision of this Agreement. Licensee shall then return to Esri the Data and Related

MAKING
SPATIAL
DECISIONS
USING GIS

*Data license
agreement*

Materials. The parties hereby agree that all provisions that operate to protect the rights of Esri and its Licensors shall remain in force should breach occur.

Disclaimer of Warranty: The Data and Related Materials contained herein are provided "as is," without warranty of any kind, either express or implied, including, but not limited to, the implied warranties of merchantability, fitness for a particular purpose, or noninfringement. Esri does not warrant that the Data and Related Materials will meet Licensee's needs or expectations, that the use of the Data and Related Materials will be uninterrupted, or that all nonconformities, defects, or errors can or will be corrected. Esri is not inviting reliance on the Data or Related Materials for commercial planning or analysis purposes, and Licensee should always check actual data.

Data Disclaimer: The Data used herein has been derived from actual spatial or tabular information. In some cases, Esri has manipulated and applied certain assumptions, analyses, and opinions to the Data solely for educational training purposes. Assumptions, analyses, opinions applied, and actual outcomes may vary. Again, Esri is not inviting reliance on this Data, and the Licensee should always verify actual Data and exercise their own professional judgment when interpreting any outcomes.

Limitation of Liability: Esri shall not be liable for direct, indirect, special, incidental, or consequential damages related to Licensee's use of the Data and Related Materials, even if Esri is advised of the possibility of such damage.

No Implied Waivers: No failure or delay by Esri or its Licensors in enforcing any right or remedy under this Agreement shall be construed as a waiver of any future or other exercise of such right or remedy by Esri or its Licensors.

Order for Precedence: Any conflict between the terms of this Agreement and any FAR, DFAR, purchase order, or other terms shall be resolved in favor of the terms expressed in this Agreement, subject to the government's minimum rights unless agreed otherwise.

Export Regulation: Licensee acknowledges that this Agreement and the performance thereof are subject to compliance with any and all applicable United States laws, regulations, or orders relating to the export of data thereto. Licensee agrees to comply with all laws, regulations, and orders of the United States in regard to any export of such technical data.

Severability: If any provision(s) of this Agreement shall be held to be invalid, illegal, or unenforceable by a court or other tribunal of competent jurisdiction, the validity, legality, and enforceability of the remaining provisions shall not in any way be affected or impaired thereby.

Governing Law: This Agreement, entered into in the County of San Bernardino, shall be construed and enforced in accordance with and be governed by the laws of the United States of America and the State of California without reference to conflict of laws principles. The parties hereby consent to the personal jurisdiction of the courts of this county and waive their rights to change venue.

Entire Agreement: The parties agree that this Agreement constitutes the sole and entire agreement of the parties as to the matter set forth herein and supersedes any previous agreements, understandings, and arrangements between the parties relating hereto.

MAKING
SPATIAL
DECISIONS
USING GIS

*Data license
agreement*

Installing the data and resources

Making Spatial Decisions Using GIS: A Workbook comes with a DVD at the back of the book labeled Data and Resources. This DVD contains the GIS data and other documents you will need to complete the projects. Installation of this DVD requires approximately 1.06 gigabytes of disk space.

MAKING

SPATIAL

DECISIONS

USING GIS

Installing the data

and resources

Follow the steps below to install the files. Do not copy the files directly from the DVD to your hard drive. A direct file copy does not remove write-protection from the files, and this causes data editing steps in the projects not to work.

1. Put the Data and Resources disk in your computer's DVD drive. A start-up screen will appear. If your auto-run is disabled, navigate to the contents of your DVD drive and double-click the Setup.exe file to begin.

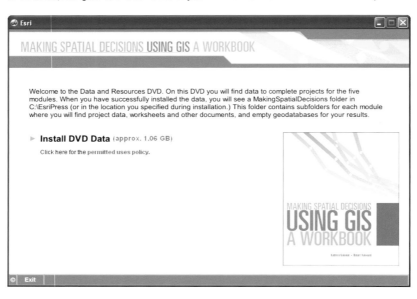

2. Read the welcome, and then click the Install exercise data link. This launches the Setup wizard.

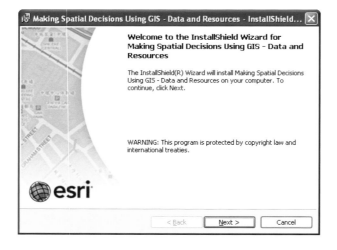

3. Click Next. Read and accept the license agreement terms, and then click next.

4. Accept the default installation folder, or click Browse and navigate to the drive or folder location where you want to install the data. If you choose an alternate location, please make note of it as the book's exercises direct you to C:\EsriPress\MakingSpatialDecisions.

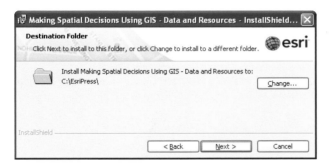

5. Click Next. The installation will take some time. When the installation is finished, you see the following message:

6. Click Finish. The exercise data is installed on your computer in a folder called MakingSpatialDecisions.

If you have a licensed copy of ArcGIS Desktop 10 (ArcView, ArcEditor, or ArcInfo license) installed on your computer, you are ready to start the exercises. Otherwise, follow the instructions for downloading and installing the trial software.

Uninstalling the data and resources

To uninstall the data and resources from your computer, open your operating system's control panel and double-click the Add/Remove Programs icon. In the Add/Remove Programs dialog box, select the following entry and follow the prompts to remove it:

- Making Spatial Decisions Using GIS—Data and Resources

Installing the trial software

A 180-day trial version of ArcGIS Desktop 10, ArcEditor license (single use) software can be downloaded at http://www.esri.com/180daytrial. Use the code printed on the inside back cover of this book to access the download site, and follow the on-screen instructions to download and register the software.

Please note that the 180-day trial of ArcGIS Desktop 10 is limited to one-time use only for each software workbook. In addition, the ArcGIS Desktop 10 software trial is only applicable to new, unused software workbooks. The software trial cannot be reused, reinstalled, nor can the time limit on the software trial be extended.

The ArcGIS Desktop 10 software installation includes three ArcGIS Desktop extension products used in this book: ArcGIS 3D Analyst, ArcGIS Network Analyst, and ArcGIS Spatial Analyst. ArcGIS 3D Analyst includes the ArcScene and ArcGlobe applications, which are used for 3D visualization and exploration of geographic data. ArcGIS Network Analyst and ArcGIS Spatial Analyst provide tools for specialized analysis tasks.

Once you have installed the trial software, go to the ArcGIS Resource Center at http://resources.arcgis.com/content/patches-and-service-packs, click ArcGIS Desktop and download ArcGIS 10 (Desktop, Engine, Server) Service Pack 1, which contains performance improvements and maintenance fixes.

All the help and resources needed to get up and running with the trial software are provided here through videos, instructions, and commonly asked questions: http://www.esri.com/evalhelp.

Uninstalling the software

To uninstall the software from your computer, open your operating system's control panel and double-click the Add/Remove Programs icon. In the Add/Remove Program dialog box, select the following entry and follow the prompts to remove it:

- ArcGIS Desktop 10

MAKING
SPATIAL
DECISIONS
USING GIS

*Installing the
trial software*

Related titles from Esri Press

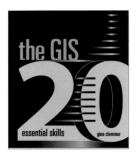

The GIS 20: Essential Skills
ISBN: 978-1-58948-256-2
The GIS 20: Essential Skills is an easy-to-understand guide that emphasizes the top twenty skills most people need to master to be successful using GIS. A quick and comprehensive introduction to fundamental GIS skills, this book includes a data CD for completing the exercises. Written for professionals with no time for classroom training, *The GIS 20* can be used for independent study, or an as-needed reference.

GIS Tutorial 2: Spatial Analysis Workbook, Second Edition
ISBN: 978-1-58948-258-6
GIS Tutorial 2: Spatial Analysis Workbook, second edition offers hands-on exercises to help GIS users at the intermediate level continue to build problem-solving and analysis skills. Inspired by the Esri Guide to GIS Analysis book series, *GIS Tutorial 2* provides a system for GIS users to develop proficiency in various spatial analysis methods, including location analysis; change over time, location, and value comparisons; geographic distribution; pattern analysis; and cluster identification.

GIS Tutorial 3: Advanced Workbook
ISBN: 978-1-58948-207-4
GIS Tutorial 3: Advanced Workbook explores the full breadth of ArcGIS software by demonstrating the complex capabilities of the tools available with the higher license levels of ArcGIS Desktop. Whether used for independent study or in a college course, this workbook will help develop advanced GIS skills. *GIS Tutorial 3* is divided into the following four sections: geodatabase framework design, data creation and management, workflow optimization, and labeling and symbolizing. Each section contains tutorials composed of step-by step instructions on how to perform pertinent tasks, exercises to reinforce the concepts taught, reviews, study questions, and real-world applications.

Esri Press publishes books about the science, application, and technology of GIS. Ask for these titles at your local bookstore or order by calling 1-800-447-9778. You can also read book descriptions, read reviews, and shop online at www.esri.com/esripress. Outside the United States, visit our website at www.esri.com/esripressorders for a full list of book distributors and their territories.